Bosc D'Antic
on
GLASSMAKING

*Including
essays on faience and
the assaying of ores*

Published 1758–80

Translated by
Michael Cable

THE SOCIETY OF GLASS TECHNOLOGY
2003

Bosc D'Antic on Glassmaking
The second in an eight volume series on how the understanding of glassmaking advanced over the course of four centuries from the early 1600s to around 1926.
Volume 1. The art of glass by Christopher Merrett (1662)
Volume 2. Bosc D'Antic on glass making (1758–80)
Volume 3. Early 19th century glass technology in Austria and Germany (1820–37)
Volume 4. Apsley Pellatt on glass making (1820–49)
Volume 5. Georges Bontemps on glass making (1868)
Volume 6. Chemical Technology of Glass by Eberhard Zschimmer (1913)
Volume 7. Glass Manufacture by Walter Rosenhain. (1919)
Volume 8. A History of the Firm of Chance Brothers & Co (1926)

© Michael Cable 2003

Published by the Society of Glass Technology, 2003

The objects of the Society are to encourage and advance the study of the history, art, science, design, manufacture, after treatment, distribution and use of glass of any and every kind. These aims are furthered by meetings, publications, the maintenance of a library and the promotion of association with other interested persons and organizations.

Society of Glass Technology
9 Churchill Way
Chapeltown
Sheffield S35 2PY

Registered Charity No. 237438
Web site: http://www.sgt.org

ISBN 978-1-917088-01-5

Bosc D'Antic
on
GLASSMAKING

CONTENTS

Introduction	1
Preliminary discourse	13
Memoir that took the prize on glass making (1760, 1780)	27
On the antiquity of glass making. The discovery of glass (A)	32
Progress in the art of glass making in England (B)	35
Glass making in France	37
Progress in the art in France since 1760 (C)	39
On furnaces and pots	41
The nature of clay used for furnaces and pot making (D, E)	44
The blending of clays, use of grog	46
New material to be used in building furnaces (F)	49
Disadvantages of coarse grog (G)	52
Bad effects of tear drops in furnaces (H)	52
Building the furnace	52
English coal-fired furnaces (I)	58
Concerning pots	58
The English shape of pot (L)	60
Concerning the materials to be converted into glass	60
Fusibility of sands. Conjectures on the origin of quartz (M)	62
Fluxes	65
False ideas of authors on the choice of fluxes (N)	66
Red and white, lean and fat potash. Adulteration	68

of potash (O)

Soda	70
Calcination of red potash, nature of the blue in white potash, Decomposition of potash by dissolution or calcination (P)	74
New theory of Prussian blue (P)	76
Purification of alkali	78
Concerning manganese, its nature and colouring constituent (Q)	80
On lime	82
On cullet	82
Calcining cullet is prejudicial (R)	83
Decolorizing with manganese	84
Batch compositions and glass melting	85
Concerning batches. Effects of lime. Glass porcelain (S)	87
Fusible spar substituted for lime	89
Liquor of flints, false explanations	90
Mixtures producing a homogeneous material	91
Fritting	93
Animal glass, the fourth type (T)	94
Melting of the batch	96
Fuels for glass making	98
Annealing	100
Important observations on annealing (U)	101
The nature of glass and the vitrifying agent (W)	103
Waste of labour and heat in the glasshouse	107
The art of glass provides the true principles (X)	108

Memoir on the cause of bubbles in glass (1758)	113
Memoir on the nature of the electrical fluid (1762)	123
Memoir on smears in glass (1765)	131
Observations on the art of Faience (1769)	143
On crucibles of the Auvergne (1771)	153
The false emerald of the Auvergne (1771)	161
Manufacture of and commerce in potash (1775)	171
Manufacture of sheet glass by the Bohemian method (1775)	181
On manufactures using fire (1775)	191
On the assaying of ores (1775)	201
Experiments on selenitic and fusible spars (1776)	219
On the evaporation of water thrown on glass (1778)	225
Bibliography	229
Plants used to provide alkali	233
Index	235

List of figures

1. Plan of wood fired furnace — 54
2. Plan of coal fired furnace — 56
3. Vertical section through furnace — 57
4. Plan of alkali workshop — 73
5. Plan of calcining furnace — 79
6. Plan of flattening furnace — 187

Bosc D'Antic: Collected works

Contents of the original volumes

Articles not included in this volume are shown in italics

	Page
Volume 1	
Memoir on the cause of bubbles in glass	1
Memoir on the cause of blow holes in metals	*21*
Memoir that carried off the prize on glassmaking	50
Notes on the above	153
A. Discovery of glass &c.	153
B. Progress in England	157
C. Progress in France since 1760	163
D,E. Nature of clay used for furnaces and pots	166
F. Composition of earths & new material	175
G. Inconvenience of coarse grog	179
H. Bad effects of drops in the furnace	180
I. Coal-fired English furnaces & pots	181
L. On crucibles. Shape of English pots	182
M. Fusibility of different types of sand	183
N. False ideas on the choice of fluxes	190
O. Potash: red, white, fat, lean	192
P. Calcination of red potash; Prussian blue	198
Q. Conjectures on manganese	208

R. Practice of calcining cullet prejudicial 213
S. Glass compositions, effects of lime 215
T. Animal glass, four types 231
U. Some observations on annealing 237
W. On the nature of glass & the vitrifying principle 241
X. Glassmaking provides the true principles 251

Observations on making faience 258

Memoir on the electrical fluid 282

Volume 2

False emerald of the Auvergne 1

Hot springs of Chaudes-Aigues 20

On crucibles of the Auvergne 31

Assaying ores by fire 51

Letter on the cause of asphyxiation 96

The inconveniences of kitchen vessels 105

Letter on claimed canard chats 124

Critical examination of selenitic and vitreous spars 126

Manufacture of potash 138

Manufacture of Bohemian sheet glass 162

Simple means of classifying irons 181

Memoir on causes of the plague 192

Manufactures using fire 236

Means of improving the commerce of Bordeaux 259

On the evaporation of water cast on glass 272

Two memoirs of Bergman; fixed air and chemical affinities	279
The art of curing hernias	309
Different states of acid in animal economy	357
Memoir on smears in glass	417
Table of contents	445
Approbation	469

INTRODUCTION

THIS is the second of three [now eight] volumes intended show the advance of understanding of glass making, especially its chemistry, since the seventeenth century. The first, *The Art of Glass* by Antonio Neri, originally published in Italian in 1612 and reprinted in 1661, was published in an English translation (with an extensive commentary that almost doubled its length) by Christopher Merrett in 1662. That became the most widely translated and frequently reprinted of all books on glass making, appearing in at least six languages and twenty editions over almost two centuries. It was reprinted as the first of this series two years ago. It gives numerous recipes for making coloured glasses but clearly shows how little scientific, particularly chemical, understanding was possible at that time.

This second volume of papers written in French a century later (1758–80) demonstrates a rather wider basis of careful observation and recording of fact about glass melting; also a considerable advance in applying scientific method. However, it also shows how convoluted attempts to understand chemical processes had become when the classical four element theory of matter and the role of *phlogiston* were still accepted but becoming difficult to uphold and soon to be abandoned.

The third contains fascinating but strangely neglected papers written in German between 1820 and 1837, by which time many of our modern basic ideas about inorganic chemistry had become widely accepted. However, the structures of atoms had not been worked out and invention of the periodic table, which makes sense of the relations between structure and properties, had to wait another half century. Those papers show that considerable progress had been made with the elapse of another half-century. Indeed understanding had by then advanced so much that the occasional serious misconceptions of those authors come as quite a shock.

Several important accounts of glass making were published during the eighteenth century, the two most notable being the famous Diderot *Encyclopédie* and the *Encyclopédie Méthodique*. However, Paul Bosc D'Antic is the liveliest and most readable of the writers on glass making of that era. Piganiol, in *Le Verre, son Histoire, sa Fabrication* (1965), considered that D'Antic left us the most illuminating of all accounts of glass making of his time. However, he was controversial and notorious (not so adverse a term in French as in English) during his life and has remained so ever since, chiefly because of his experiences at Saint-Gobain and what he wrote about them.

During the twentieth century he was called "a remarkable man of science" by James Barrelet, author of *La Verrerie en France* (1953) but also accused by Henri de Coquereaumont, author of a detailed unpublished history of Saint-Gobain (*ca.* 1920), in the following terms: "D'Antic, who had completed the deleterious work of his four predecessors and put the manufactory in a deplorable state, was dismissed in 1758." On the other hand, Piganiol spoke of D'Antic as a man of complex personality, not deficient in intelligent ideas but lacking the human qualities necessary to a successful manager who may therefore have been unable to put his ideas into practice. Piganiol also said that "It would have been impossible for him to achieve in two years the miracles expected". Piganiol implied that D'Antic may have exaggerated his knowledge and experience to gain his appointment at Saint-Gobain. Bosc D'Antic's writings need to be read with some caution but more on account of the unsatisfactory state of science at that time than any defects of the author's personality. There can be no doubt of his fascination with glass-making.

What makes D'Antic's writings so different from other writers on glass making of his era is that he clearly wishes to tell us of his own experiences and opinions not just what was generally accepted. He was evidently a colourful character, vain, ambitious, and not above stretching the facts occasionally but also sometimes admitting when he had been wrong. Those qualities do not prevent an author writing an interesting book. They may, perhaps, also indicate why his various brief biographies show gaps and inconsistencies.

His character seems to have been very like that of one of the eminent scientists and entrepreneurs of the following generation, Benjamin Thompson of Massachusetts (1753–1814), who also liked to embellish his pedigree and experience. Thompson began as a spy for the British at the time of the American War of Independence, which resulted in him coming to England, where he played a part in the founding of the Royal Institution. He then moved to Bavaria where he became very influential at Court and, although ennobled as Count Rumford, also became unpopular with leading figures there because of his grip on parts of the economy and thus on the possibilities of others making profits. His invention of the science of photometry was driven by the practical desire to find the cheapest method of illuminating workshops so that work could continue after dark. Finally he moved to Paris where he married Lavoisier's widow and eventually disappeared without trace.

Brief reviews of D'Antic's life, his connexions with Saint-Gobain, and the state of chemistry towards the end of the eighteenth century may help appreciation of his writings.

Bosc D'Antic's life

The most recent and thoroughly researched account is that given by Hamon & Perrin in *Au cœur du XVIIIe siècle industriel* (1993), a detailed history of the village of Saint-Gobain, which contains much information unknown to previous biographers. Paul Bosc D'Antic came from a Protestant family living in the commune of Viane, Tarn, about 60 km east of Toulouse. His father and grandfather were surgeons and like them Paul, born on 8 July 1726, studied medicine at the University of Montpellier but, being a Protestant, could not graduate there. He was next recorded as being a pastor in Lausanne where he was ordained in October 1751. In 1753 he had moved to Paris and was evangelizing in the Île-de-France. Until then he had been without paper qualification but received his physician's Diploma from Harderwijk in Holland on 12 April 1753 only three days after registering as a candidate; he may then have stayed there for about two years. It may be recalled that in those days medicine was the only respectable profession that offered any training in observation and deduction then taking actions based on them: many of the early pioneer scientists, especially in chemistry, were medical men.

At some stage, presumably whilst in the Netherlands, he became interested in applied science and high temperature industries. On returning to Paris in 1755 he studied with two of France's leading natural philosophers, Ferchault de Réaumur and the *abbé* Nollet, and took part in the meetings of the Academy of Sciences.

How so young a man came to be given an important post at Saint-Gobain soon thereafter remains unclear but may be hinted at by his honorary appointment as a singer at the Chapel of the Dutch Ambassador in Paris, an influential Huguenot organization. At that time the Company of Saint-Gobain was largely financed by Protestants from Geneva who frequently visited Paris; D'Antic could easily have made their acquaintance and impressed them. There is a tradition that Réaumur recommended him but it has also been suggested that he was sponsored by Turgot.

His appointment at Saint-Gobain in 1755 was made at a time when the works had been badly managed for a number of years under four predecessors (according to Coquereaumont), the last of them being Romilly. D'Antic's position at Saint-Gobain seems to have been misunderstood until recently clarified by Hamon & Perrin (1993) who showed that D'Antic was not appointed as director (works manager) but as a scientific adviser. However, Delaunay Deslandes (1723–1803), who had joined Saint-Gobain in 1752, probably after studying at Avranches, and presumably expected to become responsible for the works when Romilly left, was required to defer to D'Antic (a younger man) and follow his instructions during the two years that Bosc D'Antic spent there. Delaunay Deslandes was appointed works manager after D'Antic was dismissed and remained in that post for 31 years. He became one of the company's most notable and honoured servants but, no surprise, was very critical of D'Antic, see Barton (1989).

After being dismissed from Saint-Gobain and instituting an unsuccessful law suit against the Company, D'Antic became associated with the plate glass factory at Rouelles near Langres (to the south west of Nancy) which flourished for about twenty years despite infringing the Saint-Gobain monopoly. He left there in 1763 to run a glass works at Servin in the Doubs and at the same time was associated with a faience factory belonging to his second wife's family at Aprey 15 km from Rouelles.

In 1769 he founded a company to exploit the forest of the Margeride belonging to La Tour D'Apchier. There he set up a furnace to make flint (colourless) glass and another to make bottles. In 1771 he intended to set up a school of applied chemistry (to teach glass making) but it never opened. The glass makers had not been granted the necessary legal rights and were expelled from the Margeride: that ruined D'Antic. He then lived for some time in the Auvergne continuing his scientific studies, some of which were published.

In 1775, which was by chance the year that Dr Johnson visited France and left us an account of seeing plate glass polished and mirrors made, D'Antic was in England. He had hoped both to become a Fellow of the Royal Society (to which he read a paper on the assaying of ores in 1774) and to be given a post with the Ravenhead plate glass company but achieved neither, so he returned to Paris the next year. He later obtained a prestigious and fashionable appointment as Physician to the King and practised medicine until his death in 1784.

D'Antic was married twice, each wife having one son. His elder son, Louis-August-Guillaume (1759–1828), held a succession of Government appointments during the turbulent and dangerous years following the Revolution. He became the lover of Madame Roland and tutor to her daughter Eudora who later spurned his offer of marriage. At one stage he was arrested but released and subsequently declined to become Director of the Post Office. From 1796 to 1798 he was a Consul in the United States. He remained a student of applied science and became an acknowledged authority and prolific author on agriculture and natural history.

Saint-Gobain up to 1755

The company of Saint-Gobain had been established as the *manufacture royale des Grandes Glaces* in 1665, but had an inauspicious early history. Following several reorganizations, it in 1692 became a competitor of the existing company of the de Nehou family at Tourlaville in Normandy. At that time plate glass was made from thick-walled blown cylinders. Abraham Thévart, of the Tourlaville company, had invented the casting of flat plates of glass some years earlier but had yet to bring the process to commercial success. Initial trials were at a works in Paris, which was a very expensive place to make glass. The two organizations combined in 1695 and the new company sought another site to develop the new casting process under the direction of Louis Lucas de Nehou, nephew of the owner of Tourlaville. This needed extensive tracts of forest and abundant masonry for new buildings which might be provided by ruins. The chateau of Saint-Gobain, about 110 km northeast of Paris, met these criteria and was purchased.

Casting slabs of plate glass onto a metal table required new apparatus and skills. The most obvious feature for the glass makers being the larger scale of operations: bigger quantities of glass were to be handled and larger furnaces needed, especially for annealing of the cast plates. When all the problems had to be solved by trial and error with little previous experience to build on, it is no surprise that success was not achieved quickly. As the necessary facilities were large and expensive, crises in the financing were to be expected and did occur. These had led to the Protestant investors in Geneva becoming very prominent in the Saint-Gobain organization.

D'Antic at Saint-Gobain

In his essay of 1760 D'Antic states that a predecessor (Romilly) had claimed that a greater output and thus more profit could be attained by using pots of one third greater capacity. However, the introduction of those pots, three months later, made melting so much slower that the expected increase in output did not occur and the glass quality was much impaired. He goes on to claim that every possible reason for the problem, except the true one, was unsuccessfully put forward until he took over. D'Antic claims that he then temporarily increased the proportion of alkali in the batch to achieve easier melting until smaller pots could again be provided and the manufacture returned to its previous state.

It may be noted that using too much alkali was one of the criticisms of D'Antic made by Deslandes. D'Antic was apparently the first to extract the alkali from raw soda at Saint-Gobain, so that purer alkali could be used in the batch, but Deslandes claims that D'Antic stole that idea from him. Deslandes says that D'Antic then made glass with too much alkali and that very little of it was saleable. Deslandes claims that, on taking over himself, he steadily decreased the proportion of alkali and made much better glass. He made the notable and laudable statement that "my aim has always been to make glass with fire not salts [alkalis]": see Barton (1989).

Delaunay Deslandes also accuses D'Antic of not knowing that lime must be added, as well as sand, when purified alkali is used; also of wanting to extract the alkali present in lime for use in his batches. He says D'Antic must have seen the use of lime in Germany if he had, as he claimed, visited numerous glass works there. Yet D'Antic's comments on lime in his memoir written in 1760, although brief, are sensible. Deslandes reports that his own experiments showed one quarter as much lime as alkali salt to be best. Given the length of Deslandes service and the rapid progress in chemistry towards the end the eighteenth century, these criticisms may owe something to hindsight. It was not until about 1786 that Lavoisier recognized that melting and solidification could not be due to the evolution or absorption of a chemical substance *phlogiston* and there was still much confusion over the identification of various elements and compounds and also their properties. Correctly assigning sodium and potassium to the group of alkali metals and calcium to the somewhat different alkaline earth group was not clearly established until around 1825.

The economic and social history of Saint-Gobain has been well served by several authors particularly Hamon in *Du Soleil à la Terre: une histoire de Saint-Gobain* (1989) and Daviet in *Une multinationale Francaise. Saint-Gobain 1665–1989* (1989), as well as the exhaustive study of Hamon & Perrin, already mentioned, but the advance of technology has not been so well covered. It is to be regretted, especially by glass technologists, that the detailed history of his times at Saint-Gobain written by Delaunay Deslandes has never been published although several copies exist in archives. It seems likely to contain other interesting insights into glass making and the chemical understanding of glass makers of his time, at least as valuable as those of Bosc D'Antic.

Understanding of Chemistry 1750–80

Any translator often has to assume that he understands what was in the author's mind better than the words themselves show and thereby run the risk of introducing an error. In the latter part of the eighteenth century understanding of chemistry was improving rapidly but still very different from what it is today, so there is a distinct risk of this kind of error in the discussion of chemical phenomena in the present work. Some further comments on the chemistry of the time may thus be useful.

Since classical times, when investigation of the material world by educated people was largely a matter only of abstract thought, it had been generally accepted that all materials were comprised of four basic elements: air, earth, water, and fire. This hypothesis is generally attributed to Empedocles who lived in Sicily in the fifth century BC, who also believed that matter could be neither created nor destroyed. For many centuries practical knowledge was largely confined to useful operations such as the production and working of materials like gold, iron, and bronze, as may be seen from the famous texts of Biringuccio (1540) and Agricola (1556).

This four element model was still the basis on which D'Antic and his contemporaries tried to understand the chemical nature of their world but it may not be as ridiculous as first appears to us if these supposed elements are taken to be metaphors for the three basic physical rather than chemical states of matter that we recognize, solid, liquid, and gas, together with energy. Distinctions between physical and chemical phenomena or properties were not then clearly established.

At the beginning of the eighteenth century metallurgy was the most obvious field for studies in high temperature applied chemistry. Combustion and its role in transforming one material into another (an ore into a metal or batch into glass) was recognized as one of the most challenging problems in chemistry. Stahl (1660–1734), an authority mentioned several times by Bosc D'Antic, was the first to develop a detailed hypothesis about the action of heat in the smelting of ores and rusting of metals which led him to name the "fiery" element *phlogiston*; D'Antic also calls it the *inflammable principle*. Combustion was claimed to be the loss of this phlogiston. Stahl considered that the burning of wood or paper confirmed his view, air being merely the carrier of the phlogiston. Melting and solidification were likewise assumed to be due to the taking up or release of phlogiston. However, Boerhaave (1668–1738), who published his *Elementa Chemiae* in 1724, dissented pointing out that oxidation of metals increased their weight whereas Stahl claimed that this process also involved loss of phlogiston. Stahl had himself observed that heating lead oxide and converting it into the metal caused a loss of weight when it was supposed to take up phlogiston: so phlogiston had to have a negative mass.

Around 1780 Cavendish (1731–1810) showed that combustion of *inflammable air* (hydrogen) produced water: that seriously undermined the notion that water could be a fundamental element.

In 1774 Priestley, who had been studying gases, observed that mercury forms a red calx or oxide when heated in air but that heating the oxide in a closed container produced globules of mercury and evolved a gas (oxygen) which accelerated combustion and kept a mouse alive longer than air would have. Priestley demonstrated his experiment with mercury and its oxide to Lavoisier (1743–94), the greatest chemist of the age, whose wife was his active collaborator. Lavoisier at once realized its significance and published his own results without acknowledging Priestley. He studied combustion in detail and gave the names hydrogen and oxygen to those gases. Oxygen had been called *dephlogisticated air*; carbon dioxide was *fixed air* or *aerial acid*. On the publication of Lavoisier's *Elementary Treatise on Chemistry* in 1789 support for *phlogiston* began to fade.

Lavoisier had a modern understanding of the term element as "the simple and indivisible molecules that compose bodies" and insisted that experiments to show that they could not be further decomposed were the proper means of identifying them. He recognized the conservation of mass in chemical reactions and in his book listed thirty-three elements some of which, such as lime, were later shown to be compounds but his list included light and *caloric* which may be taken to show him on the threshold of recognizing the importance of energy. Lavoisier considered *caloric* to be "an imponderable fluid" which still retained many of the features of the discredited *phlogiston*. Nevertheless, it was an important step forward to recognize that light and energy may play important parts in chemistry.

The first quantitative atomic theory concerning the relative masses of elements and recognizing that they combined only in simple ratios was due to Dalton (1779–1844). Dalton proposed his theory in 1803 and described it in detail in 1808 but it only became generally accepted when propounded by Berzelius (1779–1848), perhaps the greatest of the Scandinavian chemists, who was Professor of Chemistry at Stockholm from 1815 to 1832. Berzelius discovered three elements (cerium, selenium, and thorium) and was the first to isolate three others (silicon, titanium, and zirconium). He also invented the modern system for writing chemical formulae. It therefore was not until about 1820 that chemists generally began to think about their subject in the terms that we understand today.

A fascinating account of the evolution of our understanding of the chemical elements and their basic properties from ancient alchemy to the modern day is to be found in Strathern's *Mendeleyev's Dream* (2000).

These major strides forward came too late to influence D'Antic's thinking. In his time few methods of analysis or chemical reagents were available and experimenters often had to rely on colour, taste, smell, and the effects of lye, sulphuric, hydrochloric, and nitric acids, as well as changes such as texture and colour caused by heating. It is clear from D'Antic's report on his examination of mineral waters that tarnishing of silver (a test for sulphides) and formation of a white opalescence on adding silver nitrate solution (a test for chloride) were well known. His attempts to understand chemical processes and the action of heat were thus hampered by both a very limited range of tools and a faulty foundation that was on the verge of collapse. His convoluted arguments trying to describe and understand various observations are thus very difficult for anyone to follow and translate or interpret today. Many examples my be found in his articles on faience and on the assaying of ores. D'Antic himself says of one problem "this has tortured my mind for a long time".

Two of the particularly trying basic chemical problems for glass makers of D'Antic's time concerned the selection, preparation, and assessment of sources of silica and alkali. As silica cannot be decomposed by heat or the common acids, it was very difficult to identify it with confidence and glass-makers tended to assume that any whitish refractory pebbles could be used. D'Antic often refers to "flint-like pebbles," *pierres du genre de caillou*, which may usually be translated as "silica": the generally accepted definition at that time was "rock or pebbles that strike sparks with steel". Many glass makers still used raw alkalis made by burning vegetation which was often deliberately adulterated by suppliers to increase its mass and their profit. Assessing the quality and composition of their alkali was very difficult; taste was often used. Some further insight into this problem is given in D'Antic's article on potash. Other important matters that were not understood included how oxidation or reduction of, for example dissolved iron, could alter the colour of the glass or how the interaction of iron and manganese decolorized glass although decolorizing in that way had long been widely practised.

Contents of the present volume

All of D'Antic's works on glass making or closely related matters are included. Examination of the list of contents of his original two volumes shows what has been omitted.

His *Preliminary Discourse* opens the volume and is followed by his longest work *The essay that in 1760 won the Prize ...* The extensive notes added to that for the publication of his works in 1780 are here included in the main text, generally where indicated by D'Antic: occasionally a note fits better elsewhere. The end of each note and return to the earlier text is marked by a row of asterisks. D'Antic provided titles for his notes (marked A to X) but very few headings within the text of the original essay; suitable headings have been added in various places to provide a better guide to the topics discussed. The rest of the memoirs then follow in chronological order but it should be noted that the first of those dates from 1758.

The figures have been redrawn in a style similar to the originals and the printers' ornaments are also copied from the originals. Brief interpolations not directly sanctioned by the original text are enclosed in brackets. Footnotes are generally those of D'Antic; those added by the translator will be obvious. Littré's and the Oxford dictionaries have been of considerable assistance.

The translation generally keeps as closely to the author's sentence structure and punctuation as gives a readable text. The main problem has been D'Antic's frequent use of very long sentences with multiple semi-colons, so sentences have sometimes been divided.

Advice and encouragement have been provided over a number of years by some of the translator's friends and colleagues including Dr J. L. Barton, lately Director of Saint-Gobain Recherche, M. M. Hamon, archivist at Saint-Gobain, Professor I. Freestone, Dr D. Martlew and Mr D. Moore of the Society of Glass Technology. Infelicities of translation or comprehension are nevertheless to be attributed entirely to the translator and editor whose mind, like the author's, has often been tortured.

The University of Sheffield
Michael Cable

PRELIMINARY DISCOURSE OR INTRODUCTION TO THE USEFUL ARTS

TO WORK to perfect the useful arts or the means which assure to men their necessities and luxuries is to be occupied with the good of one's fellows. It seems that those who are best prepared to feel this truth have long misunderstood it. Scholars have only been struck by this since the study of real physics has taken the place of systems and hypotheses.

This is one of the principal reasons why the majority of the useful arts have been and perhaps will long remain without a reasoned theory, given over to blind routine; however, it is not the only one.

Men have only felt it necessary to perfect the means of satisfying their needs in proportion as their needs have multiplied and as their luxury has led to new ones.

The people who are most occupied with the useful arts, who are the most disposed to follow the operations, are usually the least appropriate to improve them. They lack the insight necessary to develop their natural laws and establish their true principles.

Scientists are too rarely inclined to see and to follow the full scale operations of the useful arts to be able to contribute efficaciously to their perfection. Results obtained in a study or an ordinary laboratory are often misleading and nearly always imperfect expressions of those obtained in the workshops themselves.

One would be less astonished that the useful arts have made so little progress if one had given attention to the fact that three things are equally important to bring them to perfection.
1. A knowledge as perfect as it is possible to achieve of the principles appropriate to each of the arts.
2. A clear idea of all the circumstances of their operations.
3. A talent or aptitude for applying the principles to the operations.

It is much easier to acquire knowledge of the principles than of the circumstances of their operations or then to make a happy application of the principles. The true physicist or chemist possesses the principles but he commonly ignores the procedures used in practice. Very few practitioners know the principles and the majority have only vague ideas about the details of the operations. Then there are those who see the operations every day but never observe them. The aptitude for successfully applying the principles to the operations is a talent given to us by nature or by very long practice.

A work on all the useful arts which united these conditions in the best order and with the greatest precision would be a treasure precious to mankind. I do not know any such work and one could hardly expect it from the researches and toils of one man. Would not such an enterprise be beyond his resources?

Two scientists of rare merit, Réaumur and Shaw, have had the right idea, yet however deep their insights were, they have only been able to carry out quite small parts.

A body of scientists of the first order, the Royal Academy of Sciences, has adopted and decided to execute the plan of Réaumur. Nothing more fortunate for the useful arts could occur. The arts that have already been published give the highest opinion of the work and the greatest impatience to have it completed. Without lacking any respect which acquaintance with this illustrious body has inspired in me for it, even when justice does not compel me, I shall permit myself a few slight observations on its description of the arts.

The arts which the academicians themselves have been moved to study in great detail, to observe for themselves exactly and repeatedly, the design and construction, the mechanics, the raw materials together with their preparation and mixing, all the operations, methods of manufacture and products, have all assuredly been made by the hands of masters and leave nothing to be desired.

Would that the same advantages also applied to those in which they were compelled to rely on memoirs which were sent to them. One can scarcely suppose that their correspondents had powers of observations as keen and experienced as their own. Some of these correspondents lack the necessary insight; many do not possess the talent for careful observation and the majority were not born with the talent for successfully applying the principles to the operations, or have not achieved, from long experience, the facility of clarifying practice in the light of a reasoned theory. It is thus at least to be feared that the arts of this second class have had slip into them some prejudices and some errors which are all the more dangerous for having, contrary to their intentions, received the seal of truth from the Academy.

We possess a large number of works on some of the arts but it seems to me that we have few from which one can obtain a clear and precise description of the art and far fewer in which one can find any useful views on improving them. Of what advantage to the arts of wrought iron and copper founding can be the treatises of Swedenborg on copper and iron? The grossest errors are there given as incontestable truths and the most puerile prejudices of the workmen as important precepts, &c.

If the useful arts have not made great progress and are still imperfect, it is because their study is very difficult, because they presume talents, insight and dispositions which are very rare and that they require sacrifices which few people have the courage to make.

I would, without doubt, have carried this study to the highest degree of perfection if the fortunate circumstances in which I found myself, the stubborn persistence that I have shown, the passion for the objective with which I have always been animated, a sojourn of almost thirty years in high temperature manufacturing, had been supplemented by the great talents that it requires.

Although my long experience has not produced all the fruits that I would have desired, I believe that I am able to give important advice to those who, by circumstance or by inclination, are destined to undertake this work.

1. Our first care should be to nourish their spirit with the true principles of experimental physics, chemistry, and mineralogy; those of the first of these sciences will make them able to design and execute all the machines that the useful arts employ. They will find everything that they require in this respect in the works of Dr Désaguliers, Belidor, the *abbé* Nollet, etc. The true principles of chemistry will give them a proper idea of the nature and action of heat, of the manner in which it should be applied and directed, of the principles of constructing furnaces, the preparation and mixing of raw material, and of the manufacture of goods. They can draw these from the works of the celebrated Pott, most of which have been translated by Machy, and in the *Elements of Chemistry* of the eminent chemists of Dijon, Morveau, Maret and Durande, &c. &c.

Mineralogy will make them understand the nature, formation and decomposition both natural and artificial, of the raw materials used in the useful arts. I cannot, for this purpose, suggest to them a purer and more abundant source than the works of the knowledgeable and indefatigable Romé de l'Isle, &c. &c.

It is of the greatest importance that those who give themselves up to the preliminary study of these sciences should continually be on their guard against the prejudices and even the most merited reputation of any master; to become accustomed, in good time, to sacrificing the brilliant to the solidly established, the marvellous to the true. If they carry over into manufacture systems and hypotheses instead of the true principles, they will assuredly not go very far in the careers on which they have entered. We must not deceive ourselves: systems and hypotheses are no less insurmountable obstacles to progress in the study of the useful arts than the useful arts are infallibly the shipwreck of systems and hypotheses. If it were necessary we could confirm these observations with numerous examples. However, our pupils are in much more advantageous circumstances than those in which we found ourselves. It is very much easier for them to escape prejudices and false inspirations.

Our predecessors could not be guided by any completely reliable guide. The true principles of the chemistry were not sufficiently well known or as well developed as they are today. They only had largely false ideas on mineralogy. During the last thirty years these two sciences have made such progress that it may be called prodigious.

2. It would be a dangerous error to imagine that one can learn about the details of the operations of the arts and to apply the true principles to them with success by experiments in an ordinary laboratory. They are not sufficiently sensitive. It is, unfortunately, only too easily proved that the results obtained on a large scale nearly always match very imperfectly those obtained on a small scale. The art of converting wrought iron into steel and of softening cast iron according to the celebrated Réaumur has been regarded as a very important work and, in many ways, it is; nothing essential appears to have been forgotten. This great man appears to have worked unceasingly, the torch of experience in his hand. However, his researches and his advice have not led to anyone being able to convert French iron into good steel or to usefully soften cast iron on a large scale. All those who have attempted it, since him, have been ruined. The reason is obvious: Réaumur only saw the working of iron in his study; he never studied the art of foundries in the foundries themselves.

But we must not dwell on a method the faults and dangers of which I have so often demonstrated. The most clear sighted physicists and chemists now only talk, in their lectures, of large scale manufacture with an appropriate degree of reserve. It is easy to feel that it is only in the workshops themselves that one can fruitfully study the conditions of their operations. The processes are there conducted on so large a scale that the reasons for the phenomena which underlie them, & of which they are the result, are only sooner or later perceived by an attentive and well practised eye.

I say sooner or later because it is rare not to be, for a long time, the plaything of misleading appearances before coming to the desired objective. The phenomena of the arts are very numerous and generally difficult to explain because the cause is rarely simple. How many things are to be observed, considered and compared at very different times and in different circumstances! Nothing in a manufacturing process is of only trivial consequence; the raw materials, their different types, their preparation and mixing demand no less attention than the equipment, combustions, fusions, &c. &c. The mere description of the operations is immense. To save time can you refer to the workmen? Without the slightest doubt your mind will thereby be filled with monstrous errors and puerile prejudices. It is necessary to encourage them to talk frequently but without ever consulting them. Misfortune will come to anyone who delivers himself to their guidance; it is rare for his confidence in them not to cause the ruin of the establishment of which he is proprietor or has been given into his care. A complete knowledge of the conditions affecting the operations can only be the fruit of time, patience put to every test, stubborn persistence and observation as varied as sustained over many years.

One could, without doubt, save time if large scale experiments were not very expensive and if it were not of the greatest importance to avoid upsetting the established procedures of manufacture; it is only after the most mature reflexion that one dare permit innovations or decide on changes; the least serious in appearance often have the most ruinous results: one striking example will serve to prove it.

A director of the famous mirror-glass factory at Saint-Gobain, probably possessed of more zeal than insight, proposed to his colleagues at the end of 1751 that they should make *four* casts from each melt instead of the three that they had made up to that time: those concerned received with rapture the proposition of their director and pressed him to make them able to enjoy as soon as possible the considerable advantages that his discovery appeared to promise them. They only had to wait the time that it took to make the pots of the capacity necessary for the intended increase in the number of castings.

The director kept his word. He made four casts from each melt, that is to say twenty four more pieces of plate a week. One cannot fail to appreciate the advantage of this: However, the first charge required twice as long for melting and refining as had been the case with melts for three casts, altogether seventy to seventy two hours instead of twenty four. This difference should, at least, have raised questions about the claimed discovery: it did not arouse any and the result was simply attributed to the negligence of the workmen.

The following meltings and refinings required no less time than the first. It was not difficult to find spurious reasons: defective drying of the wood, a change in the frit, poorer quality raw materials, particularly the soda, were all put forward in turn.

New reasons for unease but they did not lead to the true source of the problem. At the end of ten or twelve days the plates were taken from the annealing furnaces; they were found to be of a disagreeable green-brown colour, badly refined and infected with small stones or unmelted grains of sand.

Five or six months after their manufacture these blanks arrived in Paris. Hardly any survived grinding and polishing because of the small stones. Those that did escape from these two operations could not be sold and remained in the warehouse because of their brown colour, bad refining, and the sand distributed throughout them.

This was not the only problem discovered by those involved. The four casts not only produced more bad glass than the good glass that had been produced by three casts, but because of the excessive melting and refining times, they were scarcely able to make more than two melts a week.

The pots, necessarily more worn by refining for three times twenty hours and by the weight of a quarter more raw material [in fact a third], had a much shorter life. One of these pots which lasted fifteen days was regarded as a rare phenomenon: when working with three castings there had been none which could not last at least a month and a half.

It can be seen that there was an enormous loss of glass and that the furnace, being continually flooded with vitrified materials, was much more rapidly worn out.

Those concerned in this manufacture made vain efforts to re-establish good production over a period of three years and, during that period, had to face the impossibility of making any saleable cast plate glass. They would have been ruined beyond recall if they had not found in their warehouse in Paris rejects of old plate glass sufficiently thick to have their defects removed by a new grinding and polishing. This old stock in the warehouse permitted them to maintain their sales and meet their expenses.

At the moment when this precious stock was exhausted, towards the end of 1754, one of the most clear sighted and respected of those concerned did me the honour of consulting me and proposed that I should make a journey with him to Saint-Gobain.

Having arrived at the factory, I was not long in discovering that the too great capacity inconsiderately given to the pots was the sole cause of all the problems which they demonstrated again and again. It is evident that by this change they had destroyed the proper proportion that must exist between the furnace and the pots, that the degree of heat in the furnace which was sufficient to melt and refine, in about twenty four hours, 4800 pounds weight of glass was insufficient to found and refine properly 6400 pounds in the same time.

The reason for the bad production being clearly established, it was easy to find the remedy; it was only necessary to return things to the previous footing, but to make pots of the old type and to dry them properly was not merely a day's work. The drying of large pots requires at least three months, which would have been a considerable period of profitless expenses. I was able to save the directors from them.

By adding 30 pounds of fixed alkali of soda to each frit of 500 pounds, the fabrication changed completely: the duration of melting and refining was only twenty four hours; there were no longer any stones in the cast plates; the glass was much better refined and of a less disagreeable colour: I had predicted these precious advantages and they were the inevitable result of increasing the flux.

It may be felt that too great a quantity of flux was used in this batch but this was a disadvantage temporarily necessary to destroy a much greater evil.

This example will doubtless inspire a useful suspicion, a wise restraint, over innovations most likely to be advantageous to the directors of the enterprise.

3. The arts show more or less marked relations with each other so that they can assist each other, but glass making is the basis of nearly all the others, particularly metallurgy and pottery, and all its numerous branches deserve very careful attention. However important this point may be, we shall confine ourselves to advising our pupils, if they have the choice, preferably to begin by studying the art of glass making. The main reasons for this advice are developed in the first volume of this collection.

4. The study of the useful arts has its charms but also its thorns. It is surely not agreeable to pass one's life in the middle of the woods, with coarse and very ignorant men; often to be obliged to travel in rain and snow; to be the first to rise and the last to bed, necessarily to rise and go out at any hour of the night; to be continually apprehensive and ceaselessly in fear of mishaps which occur only too frequently especially in industries relying on fire, &c. &c. An iron determination, an unusual courage, a passion for the subject, can alone triumph over all these distasteful aspects.

However, we must disguise nothing from our pupils; they would have good reason to reproach us if we did not forewarn them of all that can happen to them.

Riches do not always accompany talent and insight, it is likewise rare that those people who appear born to improve the arts are sufficiently rich to become proprietors of factories.

These establishments are usually very considerable enterprises too large for one man to be able, or to wish, to provide all the funds and run all the risks. It is usual to form associations or companies to undertake these large scale exploitations.

New subjects for distaste, new sources of trouble and grief arise for those who devote themselves to the useful arts. If they are not sufficiently favoured by fortune to establish factories for their own profit, they will obviously be obliged to set them up and direct them for the benefit of others, or to associate themselves with other people.

Their favourite passion fixing their attention entirely on the interests of the arts, they necessarily neglect the interests of their liberty, their reputation and their fortune; they will pass lightly over the conditions of their employment, not going deeply into any details, judging people favourably on the superficial appearance of honesty; they will take the affected politeness of their associates or colleagues for evidence of esteem and confidence; they will not perceive that the association has often not been formed for the enterprise but rather to provide the means of having a large fund of cash and of abusing it; not thinking that the enterprise is sometimes ruined before it has begun; incapable of deception themselves they do not imagine that others can be different, &c. &c.

How I complain on behalf of the students at this point possessed by a love of the arts! They will undoubtedly sooner or later be the unfortunate victims of ingratitude, stupid vanity, cupidity, bad faith and injustice by business men.

The majority of financiers attach a much greater value to money than to talent and insight, never cease to contradict the director of the enterprise, take pleasure in mortifying him and only take notice of him with affected consideration. It is rare for business associations to recognize the sacred nature of their undertakings and even rarer that they fulfil them to the letter. If they could confine themselves to forgetting expectation of benefits, students of the arts could console themselves; but often the most fortunate efforts, the most precious discoveries, the most numerous and useful services are evidence dangerous to themselves and become a perpetual source of the most flagrant injustice. These bodies suffer from the stupid vanity of not wishing to appear to owe their success to one man; so they will retain him only as long as they believe they have an indispensable need of his services; do they imagine that the sources of his talents and his insights are exhausted? If he does not grovel at their feet, dares to place a value on his services, claims fulfilment of the promises made to him, invokes the undertakings made with him, his destruction is secretly sworn; there is no type of persecution that they will not use to force his retirement.

Whilst the director seeks day and night for new ways to make the enterprise more flourishing, the company opposes his intentions, incites his subordinates against him, encourages the workmen to fail him, interrupts the payment of the running expenses or allows them to languish, writes to him in a disagreeable manner, treats as ruinous fancies his most useful and most soundly established discoveries, announces enormous losses to him when the profits are as real as they are large; make him take responsibility for events which he could neither prevent nor foresee. They cover him with all kinds of ridicule in public, credit him with all kinds of absurdities; misrepresent his mind to cause suspicions that he lost it and, to make his own justification very difficult, give him an excessively difficult character, incapable of all society; spare no care, no effort, to give unfavorable impressions of him to the Government and if the Minister is not as fair as he is far-sighted they will heap injustice to him and seek to excuse it by legal authority. If the enterprise fails for reasons unconnected with the director of the works, on the commercial side, by bad administration of the funds, by intrigue or malpractice of one or more of the members of the company, they will still attribute the cause to him and no less regard him the author of its ruin. They will surely not neglect to engage him, in his state of ignorance and without any ability, in matters of commerce, envelop him in inextricable legal proceedings or make him play a very unjust role. Lastly, if he publishes useful discoveries and his works have been well received by the public, the hornets of the republic of letters, who unfortunately are only too common, who have nothing to draw on from their own resources, will copy him without acknowledgment and join themselves with his persecutors to complete his ruin and continue his misfortune right to the tomb.

This picture that we have just described is not overdrawn. I am, unfortunately, only too well able to prove that all these features have been faithfully reproduced. Our purpose would certainly have failed if we have made students of the useful arts discouraged in the slightest degree, or if we have formed in their minds suspicions injurious to all those involved in great enterprises. We have only intended to inspire them with prudence and enable them able to avoid shipwreck by making them understand that they should never undertake an important contract without detailed advice from a good lawyer.

We are far from believing that all commercial companies are capable of the excesses we have described. It would be easy to name many where justice and honour have never ceased to reign.

The principals of the factories have been duped much more often than the opposite has happened. Companies can easily cover themselves against this risk; they have only to avoid confusing true disciples of the arts with those who are pretentious. The former have the talents and insight as well as a well considered experience and a true love of the arts. The latter have only secrets, often imaginary, routine recipes, an uncertain practice and only love the arts for the money that they can procure. Given such striking differences who could make the wrong choice ?

5. One usually finds in large factories workmen from different provinces, sometimes from different nations, and always of different characters. Some of them are honest, calm, and reliable, but the majority are libertines, addicted to wine, muddled, intriguers, always seeking to cause disorder, and ready to run away. For them a hundred leagues on the road is no more than a short walk. Some are very intelligent and others mere robots.

How to make them comply to the same rule, and make such diverse wills co-operate for the same purpose? The art assuredly is not easy. The force of laws and the direct authority of the master can restrain the workmen but are insufficient to ensure that they do good work. It also appears to me important only to make use of these two serious sanctions in extreme cases. The true secret is to gain their respect.

It is rare that a clear sighted, firm, just and humane director does not succeed in managing the workmen very well and making them execute accurately whatever is required for the greatest good of the manufacture. He who makes a good manager is sure to be obeyed; his orders are clear, precise, unambiguous, never lead to faults, and are obviously related to the greatest advantage of the enterprise. He is inflexible about the orders that he has given, does not listen to any complaints on this essential point; does not suffer any suggestion or any action contrary to good morals; he is convinced that where morals are not respected good order cannot reign and that without good order there can never be good work; he is no less firm with his foremen than with his workmen.

He must never confuse firmness with harshness. This is as very great and very dangerous error; the former is necessary and has very desirable effects, the latter is unfair, discouraging and revolting. He should only require of those that chance has put under his control whatever is strictly within their duties and, even more important, that their health should not be impaired; respect the conditions of their employment, see that they are paid punctually, take care that their accounts are kept in good order and copies given to them whenever they wish, take up their cause in disagreements, even with their foremen; never give signs of confidence or show preferences except to those who merit them by the steadiness of their conduct and their good work.

If they are ill, if they have difficulty in subsisting because of their numerous family, if they need advice, he must set aside the manner and the language of the master and take on the feelings and courage that a tender father would devote to the succour of his children. He will console and sustain them to the limit of his ability and with all demonstrations of true affection. Such a manager will indubitably be adored by his workmen; none will dare to fail him, all will take pleasure in doing beforehand whatever will please him.

6. In our picture of the industries of the kingdom using fire, which can be seen in our second volume, we have proposed two means which would infinitely facilitate the study of the useful arts. The first consists of sending into the factories people as profoundly versed in theory as accomplished in the practice of the manufacturing process to instruct the proprietors or their directors but if one should have thrown the control of the processes into error, if the people chosen were ill-informed or should, under various pretexts, try to extract from the manufacturers details of

their compositions or their own special procedures, this method would not be of any use; it would only inspire distrust.

The second means, undoubtedly the most efficacious and the most appropriate to produce the most rapid progress in the arts, is to establish schools where the students can be instructed in both the theory and the practice, where the true principles can continually be applied to large scale operations. However, both these means depend absolutely on the government.

P. BOSC D'ANTIC

MEMOIR

which gained
in 1760,
the prize offered
by the
Royal Academy of Sciences:

*What are
the most appropriate means
to bring economy and perfection
to the glass works of
France?*

Non fingendum, aut excogitandum, sed inveniendum,
quid natura faciat aut ferat. Bacon.

1780
Paris, Rue et Hôtel Serpente

THE ART of glass making is one of the most important by which chemistry has enriched mankind. It provides us with the most agreeable and commodious receptacles. Without depriving us of light it gives us cover from the injuries of the air. The conservation of innumerable precious liquids is due to it alone. With its aid we can remedy the defects of our sight which the ravages of the years rarely fail to produce. Whence do our apartments obtain their most beautiful and most noble decorations? From the art of glass making. Few of the sciences or the arts can take place without its aid. Without it there would be no natural history, astronomy, experimental physics or, above all, chemistry!

An art of such widespread utility and possessing phenomena so apt to stimulate the imagination should already have made great progress. This consequence appears natural but the art of glass making is very imperfect. From its cradle it has been abandoned by chemists, one might say, and given over to men incapable of understanding all the resources. It degenerated into blind routine. A small number of scientists have attempted to draw it forth. Persuaded that nature reveals herself imperfectly in ordinary laboratories they have taken the trouble to study the means in glass works.[a]

I believe that *Agricola* is the first to have written in some detail on this art: but what he says about the materials from which glass is made, the furnaces in which it is made, and the manner in which it is made, are only

---

a. Georg. Ag. de Re metal, lib. 12 and de Nat. fossil. lib. 5 p 274, 275.

simple descriptions of what he had seen practised in the glass works of his day. One finds in the twelfth book of his treatise on *metals* scarcely any certain principle, any judicious comment or any useful insight. There are even important errors to which we will have occasion to return, amongst others that *rock salt* combined with sand will produce glass.

Neri is regarded as the oracle of the art of glass making. However he says not one word about furnaces or pots which is soundly based. The batch composition to which *Kunckel* devotes four lines is certainly more valuable than all that Neri says about the different ways of preparing the materials or making crystal[a]. His commentators do not offer us any greater resources. *Merrett* only adds to what Agricola wrote on furnaces and the manner of working glass a few English practices and little of consequence. His notes on Neri prove that he had read more than he had experienced himself. *Kunckel*, with less trouble, would have gone much further if he had known the principles. In his remarks every thing is reduced to a few particular methods, mostly practised in the glass works of his time and to a few useful observations. *Neri* and *Merrett* had in view only the production of the best glass without consideration of the expense. *Kunckel* considered that perfection in the art consisted of making the best quality at the lowest possible cost, which has merit. I do not believe that one can gain much insight from the summary that the celebrated *Henckel* made from Neri, Merret, and Kunckel nor from what he added on the three types of glass, mineral, vegetable and mixed[b].

The art would undoubtedly have been enriched by useful discoveries if this knowledgeable mineralogist had been able, as he said himself, to carry out operations in glass furnaces.

The art of glass making of *Haudicquer de Blancourt* is, with a few changes and additions, a translation of what Agricola had written on this subject, the seven books of Neri and the notes of Merret. This work could be read with pleasure if the author did not delve, too frequently,

---

a. See p. 101 of *The Art of Glass* (French transl. in quarto).
b. Henckel. *Flor saturn.* cap. 11.

into extraordinary ideas from alchemy.

Haudicquer arrogates all to himself and does not cite even Neri. This translation does not seem to me more likely to lead to improvements in glass making than the originals.[a]

What the celebrated *Boerhaave* has written on this subject in his *Elements of Chemistry* does not, I venture to say, correspond to his reputation; there is probably no weaker morsel in all his works than that on vitrification. See his *Elements of Chemistry* vol. VI, p.157 also vol. V, pp. 276, 277, 316 and also the French translation. We do not delude ourselves that much help is to be found in the works cited by *Merrett*[b] towards the end of his preface.

One sees in the authors we have just been discussing scarcely any principle solidly established nor any phenomenon clearly explained: everything is reduced to a few approximations, particular methods and precepts relative to the materials that they had in their hands or to the furnaces by which they were served, all of which is of little value to those operating in different circumstances. They say nothing satisfying about the material, its preparation and the construction of furnaces; on the composition and the shape of pots, on the proportion that should exist between the pots and the furnace, on the most advantageous degree of fire; on the nature of the raw materials to be converted into glass, the causes of purification, of greater or lesser transparency, of colours, of greater or lesser strength, of bubbles, cloudiness, gall blisters, tarnishing, threads or cord; on the nature of good annealing &c. Is it an art the theory of which should be so imperfect? It is to be presumed that the insights that one can acquire every day in chemistry can put us, in a little time, into a state of deeper and more extensive understanding. No one seems to me

---

a. 2 vol. duodecimo, printed in 1696 by J. Jombert, reprinted and augmented by a treatise on precious stones, and glass for mirrors, in 1718, by C. Jombert.

b. V. liber com. alch. part I cap 20. Ferrant. imper. lib. 14 and 15; and Pott, lib. 6 cap. 3, &c.

to have provided us with more or better material than the scientist Pott[a] but it is necessary to be more than a simple artist to put it into practice.

Note A

ON THE ANTIQUITY OF GLASS MAKING. CONJECTURES ON THE DISCOVERY OF GLASS

One should be even more astonished that this art has made so little progress when it has been known from the most ancient times. At the time of Strabo[b] the glass works of the great Diospolis, capital of Thebes, and those of Alexandria were very famous. It is certain that they then made cups of a glass the purity of which was like crystal, other glass vases called *alassontes* supposed to have represented figures, the colours of which changed according to the direction from which one looked at them, counterfeited precious stones and that they ground, engraved and gilded glass[c].

The scholars who appear to have worked with the greatest success to give us knowledge of the arts of ancient peoples, such as the author [de Pauw] of philosophical studies of the Egyptians and the Chinese, believe that the glass works of Egypt are more ancient than those of Tyre and Sidon.

It is not likely that these early glass makers made flat glass to any extent either by blowing or by casting plate for mirrors. The term *specula* probably standing for *specularia* which is found in Pliny[d] only seems to describe small very thick pieces of glass, usually round, mounted in

---

a. See in the Mem. of the Acad. of Berlin the memoirs of this great chemist on glass makers magnesia, on crucibles, on gall and, above all, his Lithog. Pyrothec. and others.
b. *Geograph*. Lib XVI
c. See Athénée, Lib. V, cap. VI
d. See *Nat. Hist*. book 36, chap 26

plaster for use as windows such as one finds nowadays in many parts of the Levant and Turkey.

However, one must not deny that flat glass may be very ancient. M. Soufflot, the celebrated architect gave me, some time after his journey to Italy, a fragment that he had found at Herculaneum. This glass was almost three twelfths of an inch thick, was quite wide, very transparent, of a colour tending to green and had obviously been blown because three gathers were evident.

The origin attributed by many authors to glass making has always seemed a fable to me. Merchants, it is said, having lit a fire on the river or sea shore in Phoenicia saw that the sand had melted and thus found, inadvertently, the method of making glass. One would have to be extremely credulous to believe that they saw the sand melt! It is known that a reverberatory fire is necessary to make glass. "The combination of accidents", says the knowledgable M. de P*** [de Pauw][a], "has not as much power as is commonly believed": the processes must be developed one after the other. Chance could have had some part in the invention of glass which could only have been discovered after the art of the potter: there must have been pottery and even a body approaching that of porcelain before there was glass. Many nations stopped with the discovery of porcelain without being able to go beyond that; others knew only a sort of enamel. For example, no one knew how to make glass in the entire extent of America in 1592 although there were some savages there who knew how to glaze clay pottery according to Narborough. This origin of glass-making seems much more likely to me: the art of the lime burner and the tile maker led to the idea of making glass. It is very rarely that one burns lime or fires tiles without seeing the flow of some drops of glass."

* * * * * * * * * *

Let us rapidly cast an eye over the practice of the art of glass making. Stop a moment and consider the routine production; see what is the quality and the price of glass products. In the rest of this memoir we shall have

---

a. See *Recherches philosophiques sur les Égyptiens et les Chinois*.

occasion to examine the furnaces, the pots, the compositions and the methods now in use.

There is no place where glass making had been put on so brilliant a footing as at Murano. The Venetians had a considerable commerce in mirrors, in crystal and in all types of glass. They have entirely lost this important branch. There only remains one man in Venice who makes good quality crystal and he sells it at an excessive price. The plate glass of Murano is the worst in Europe and, although rather less expensive than ours, is not sought after.

The English glasshouses have a great reputation. They are not old. Their rapid progress has been due to the singular attention of the Government not to give them impediments nor to confuse the public interest with the private. Plate glass, crystal, white flint and common glass form today a considerable branch of the commerce of Great Britain. Foreigners consume four fifths of the English plate glass. There is no country where the English do not find ways of introducing their products in crystal and glass. In other times they drew on France for nearly all the glass that they needed; today they supply us with chandeliers, lanterns, drinking glasses, optical glasses of all sizes, etc. The manufactures of London are only surpassed by those of *Neustad* for the beauty of their plate glass. One can see small pieces at the house of Mr *Sayde,* optician to the Queen, in Paris, quai des Morfondus. The large pieces are very expensive.

Plates one hundred and forty four inches high by forty inches wide are sold for up to a thousand guineas[a]. However flourishing their glass works may be, the English cannot flatter themselves, with *John Cary*, that they have been *brought to the highest perfection.* Their crystal is not of a good colour: it tends to be yellow or brown, partly because the red colour of the manganese is dominant. It is so badly annealed that it sweats salt, tarnishes rapidly and is full of seed and cloudy. A look at the slabs of crystal that the English make for optical work will be convincing. There

---

a. See the first vol. chap. 9 of the Essay on the State of Commerce of Great Britain.

is another important defect, that of being extremely soft. They sell their wares very dearly; perhaps they would be forced to lower their prices if they had some competition for their lanterns and optical glasses.

They have a considerable commerce in crystal and white flint products in many parts of Germany, in Saxony, in Bohemia, in Franconia, the Palatinate, &c. We obtain from these different places, for considerable sums, chandeliers, chimney piece handles, flasks, carafes, drinking glasses and goblets, table crystal, window glass, dials, embossed glass, blown and cast on a table, &c.[a] The most beautiful German crystal has two advantages over that of England, being white and less expensive, so that it sells in Paris for from three pounds ten shillings up to one hundred shillings a pound already cut. One may add to the defects of English crystal that it is cordy and rarely free from small stones or grains of grog. The flat glasses of Bohemia and the Palatinate are very far removed from the perfection of which I believe them capable.

They have a great number of defects but the most troublesome is to be of an unequal thickness and wavy. The cast glass of Nuremberg is usually well refined, very well polished and sell at least twenty five percent more cheaply than our own. It will be found in the workshops of many mirror makers in Paris and most of those in this province.

Note B
PROGRESS IN THE ART OF GLASS MAKING IN ENGLAND

Since 1760 the English have made considerable progress in their glass houses, as in many other branches of commerce.

Their window glass leaves nothing to be desired in the quality of the glass; it is of a good tint and well refined but these glasses, made by the crown process, have the defects inseparable from this type of manufacture,

---

a. See the decrees of the Standing Council of the glass industry.

they are sour, badly annealed and out of true. Moreover this method cannot give very large panes such as 30 × 30 or 36 × 30 [presumably inches], not only because these glasses are too thin and too warped but also because they would require disks of eight or nine feet diameter, which is impossible. The assorted English glassware, drinking glasses, carafes, etc. is white and well refined but, because of the lead used in its composition, is much too heavy.

The blown English plate glasses are of a much better colour than the cast glasses and they are also better refined. Both types of glass have seemed to me a little too fragile and lacking the substance required, in particular, for large pieces. These two defects are not easy to correct; it would take too long to explain the reasons in detail here.

The English have a very beautiful glass for chandeliers. The chandeliers that they make from it and which they polish very well, of which they cut and arrange with the greatest art all the pieces, necessarily reflect all the colours of the rainbow and exceed those of rock crystal in appearance and in their overall effects. I have seen these chandeliers sold for up to two hundred *louis d'or* (4000 francs).

The discovery of *flint glass*, which gives such astonishing effects, is entirely due to Great Britain. That which is made at present is very far from the perfection of which I believe it to be capable; one could even say that the English have lost nearly all the fruits of their discovery and for a simple reason of which they are insufficiently aware. It is very rare to find among them *flint glass* which is not infected with streaks, white specks, or cord, and is not cloudy. Although several knowledgable companies have crowned their achievements with memoirs on the manufacture of *flint glass*, it seems certain that no true flint glass is made anywhere else except in England. This is because the art does not consist only of introducing as much lead oxide as possible into the glass.

There is no country where fine white [flint] glass is as dear as in England. The raw materials, red lead and potash from Canada, of which it is composed are nevertheless cheaper than almost everywhere else. The coal used by the English to fire their furnaces does not cost them as much as the wood used by nearly all the glass works in other parts of Europe. The labour for this type of work is scarcely more expensive in England, even in London, than in our glass works. What then is the reason for the very high price of white English glass? It is the enormous tax imposed on it.

Various types of white glass such as that for chandeliers and plate for mirrors pay to the government about half of the price for which they are sold. On leaving the kingdom the Government pays back without any difficulty the tax that these goods have paid. The English appear to have a maxim that heavy taxes should be charged on all works of art that men can do without. Consequently it is only the rich men who pay the greatest imposts and without being forced to do so: this maxim conforms to humanity and to justice and cannot be harmful to industry.

It is easy to predict that the art of glass making will make more rapid progress in England than in any other part of the world. The richness of these islanders, the patriotism that they carry to the point of enthusiasm, the good quality and abundance of their lead oxide, potash and coal, the ease of transport by canals, of which they have so many, and by rivers nearly all of which have been made navigable. Above all the Society of Arts, established far above our ability to applaud it, a truly praiseworthy association in Great Britain, is ceaselessly occupied with improving the useful arts and each year dispenses for their encouragement 100,000 pounds in our money. All these inestimable advantages do not allow us to doubt that this important art will quickly be brought to the highest point of perfection by the English.

* * * * * * * * * *

GLASS MAKING IN FRANCE

The glass from our glasshouses can be divided into four types: bottle glass, common green glass called *chambourin*, white flint, and crystal. I know only three factories in France that make good bottles; Folombrey in the forest of Coucy, Anor in the French part of Hainault and Sèvres near Paris. Those that are made in Bayreuth and Delln in Brandenburg are superior to ours and sell more cheaply. Our white flint would scarcely pass for the ordinary glass in Germany and our crystal for foreign white flint. To convince oneself it is only necessary to compare fine flat glass or *double-fired* from Normandy with the common flat glass from the Palatinate and crystal from the glass works at *La Pierre* with Bohemian white flint or with the pieces of English white flint that may be seen at M. *Sayde's*. There is, I believe, no worse window glass than that made by our large factories; it is full of defects, bubbles, cord, drops, stones, badly annealed, tarnishes rapidly and is so coloured that it is hardly transparent even when very thin. In a few glassworks they have tried to imitate English lanterns but are far from achieving complete success.

At the end of the last century *Abraham Thévart* made a particularly beautiful discovery in glass making: he found how to cast plate glass for mirrors. What has been the result? Plate glass of a larger size and nothing else. The products of our glass works are of bad quality and very expensive.[a] Would one be stretching the truth to suggest that our glass works are more useful to Spain than to France? They use almost two millions of soda annually from Alicante and Carthage: but we must not dwell on a matter which is such a disgrace. No one can ignore the fact that our glass works are in a deplorable state. The extraordinary prize that the Royal Academy of Sciences has proposed to support the views of a zealous citizen do not allow us to doubt it. This respected body only promised the palm to whoever would give the most appropriate means of bringing perfection and economy to the glass works of the kingdom. This does not require only that one makes good glass, one should also be able to make it at low cost and, with these two conditions fulfilled, our glass work would certainly be able to maintain competitiveness with those of foreign countries.

---

a. Our large pieces of plate cost fifteen or sixteen francs per pound weight: the calculation is then easy. The plate is 3 to 4/12 inch thick and a cubic foot of glass weighs about 175 pounds.

Note C
PROGRESS IN THE ART OF GLASS MAKING IN FRANCE SINCE 1760

Glass making in France has changed face since the publication of my memoir. A large number of works have been set up to make white window glass and miscellaneous ware. There no longer are any French glasshouses making *chambourin* glass for rollers or medical phials. Miscellaneous products in white glass are no longer more expensive than green glass was in 1760 and table glass is considerably lower in price although potash is dearer. I can thus flatter myself to have produced, by my researches and works, an advantageous revolution in this important branch of industry and commerce.

It is notorious that the celebrated Saint-Gobain works making plate glass for mirrors had, in 1755 been unable to make any saleable cast plate glass during three consecutive years. With the intention of making a larger quantity of plate glass they had increased by about a third the capacity of the pots without increasing the capacity of the furnace. Without realizing it they had evidently destroyed the proportion required between them. It was quite impossible for the heat of the same furnace to melt properly a third more material. By increasing the fusibility of the batch, by an addition of flux, I was able to demonstrate that the source of the trouble was none other than I have just indicated. By bringing the pots back again to their proper proportion I was able to bring this manufacture back under control. I was able to do more and establish their operations on principles as simple as they were certain. It is no longer possible for them to lose their way, at least not to the extent that they had before 1755.

It is no less known that I alone set up the manufacture of plate glass at Rouelles in Burgundy; that in very little time I made plate-glass workers who were both skilled and intelligent from a large number of wood cutters and that, in setting up this establishment, I was able to make such great economies that, to the satisfaction of my associates, I was able to set prices 50% lower than the Saint-Gobain tariff for all sizes.

In my prize-winning memoir I was only able to treat the general principles of glass making and few people were in a position to make a successful application to the different branches. Partly for lack of insight and partly because the government has not given this important type of industry all the attention that it merits, importing of glass into the kingdom is still very considerable. There are few of our towns of the first or second rank where one does not find shops with cut and uncut Bohemian glass and the English continue to supply us with optical glass and crystal both colourless and coloured.

* * * * * * * * * *

I do not doubt that very many people, rendering judgement on the views and zeal of the Academy, will not put this prize on the same rank as squaring the circle or finding the philosopher's stone, &c. People are generally persuaded that it is impossible to make in France glass as good and at the same price as in Germany. In Saxony, Bohemia, Franconia, and the Palatinate good flint pebbles, they say, are very common but we lack them; fixed alkali salt, extracted from wood ash, is there abundant and at a low price but we have only inferior and very expensive raw materials; wood costs nothing there but we buy it at an excessive price. Is there no exaggeration here? Is not our vanity seeking to excuse our negligence? Given that everything appears to be just as I have said, has the matter been sufficiently closely examined? Have sufficient researches and trials been made to decide that our glass works could not find, in the kingdom, at least the equivalent of the advantages of the glass works of Germany? I give myself credit, for the good of my country, of doubting it. The favourable circumstances in which I have long found myself have allowed me to consider matters from so many points of view that I even believe that it can be proved that France could stop making imports and make here glass as good and at as low a price as that from foreign countries. It is not hollow speculation nor vague reasoning that provides these proofs; I draw only on experiences frequently repeated and on a large scale. If

I propose anything that I have not submitted to this touchstone, I will take care to say so.

ON FURNACES AND POTS

The intelligence of the founder, the keenness of the workmen, the choice and purity of the raw materials are poor resources for the master glass maker if he does not have a good melting furnace and suitable pots. However little disposed to vitrify may be the materials of which he builds the one and makes the other, his products will be infected with *drops* and *cord*. Negligence or incompetence in the preparation of these materials will produce ruinous degradation or unwanted colour in the glass. Defects in the proportions of the shape will necessarily prejudice good *refining* and increase the consumption of wood or coal. The furnaces and the pots are the most important part of glass making and, I venture to say, the least known. I have read no work from which one can gain sufficient help on the materials to be used, the manner in which they should be prepared, or the most advantageous shapes of furnaces and pots.

In most of the English and some German glass works the furnaces[a] are built from an incredibly hard stone. This is a type of sandstone called, in some parts of France *mouillasse* and in others *pierre à ouvrage* [ganister]. Furnaces built with this type of stone allow a great deal of heat to escape through the joints, consume a great deal of wood and have short lives. The most skilled mason cannot avoid the joints, make the interior as closely bonded as it should be, nor give it the most appropriate shape without impairing strength. Most of these sandstones contain a ferruginous base and vitrify, spalling or producing grains: they scarcely allow working for three or four months.

Clay is the best material for furnaces and pots. There are many types but one can employ with confidence only those designated in the mineralogy of *Wallerius* by the names white clay, grey clay, pale refractory clay,

---

a. See Pref. to Merrett p. 43 (quarto)

brown refractory clay, blackish refractory clay, porcelain clay, and pipe clay. All of these become bleached by a fierce heat. I do not consider, like *Wallerius*, that the last two are types of marl.

This knowledgeable mineralogist seems to me to have confused two very different earths because one becomes hard when fired but the other becomes chalk. A foreigner like Wallerius can say that only one type of refractory clay is found in France: but can a Frenchman like Haudicquer de Blancourt be excused for saying that the plastic clay of Belière, near Forges in Normandy, is the only one in this kingdom which does not collapse in a glass factory's heat?[b]

In the same province one can find many types of very good white, grey and blackish clays. At Montdeber not far from Nantes in Brittany one can find white, at Villentrode in Champagne grey, near Bar-sur-Aube grey, at Épinac in Burgundy brown, at Suzi in the Laone grey, in the Forez brown, at Méréviels near Montpellier white. Pure clay is assuredly very common in France: there is no master glass maker who cannot obtain it at little expense. Some investigation with a sampling probe and a little intelligence should allow him to discover it and obtain supplies. Foreigners, above all those in certain parts of Germany, may perhaps envy our advantage in this.

One never finds clay which is not mixed with some more or less dangerous foreign bodies. In some there are small pebbles,[c] in others pyrites, fossil coal, and a little sulphuric acid; in all some sand, a few roots and some iron-rich clay which is usually revealed by red or yellow streaks. Straw, sprigs of plants, pieces of wood and ordinary earth should be attributed to negligence by those who dig the clay. These heterogeneous materials are always prejudicial and sometimes fatal because they leave voids when they have been combined or dissipated by heat or when they

---

a. See the *Lithog.* of Pott and the 2nd chap of Vol. I of Geller's *Chem. metall.*
b. See p. 35 in de Blancourt, Vol. I.
c. See p14 vol. 2 *urb. hierne. Tentam. Chem.*

react with the clay during melting: such are ferruginous clay, pyrites, sulphuric acid, pebbles, and sand contaminated by glass or fixed alkali. But the most redoubtable enemy of the master glass maker is the ferruginous substance found in the clay: it is the principle cause of its fusibility and hence of drops and cords, or bad colour in the glass, and of the short a life of furnaces or pots.[d]

There is no glassworks where they do not fear the bad effects of these foreign bodies. It is in hope of preventing them that the clay is picked over with great care. It is broken into small pieces and all those that are coloured or heterogeneous are removed. However, it is easy to feel that whatever care is taken, the operation is insufficient. The sad experience that I had over many years gave me the idea of a better, more certain, and less costly method. The well-roasted clay is broken coarsely; the fragments showing most red or yellow streaks rejected and the rest put in a large wooden box at least ten inches deep which is filled two thirds full. This is then filled with hot water in winter or, if one wishes, cold in summer, so that the clay is covered by two inches of water. It is important not to disturb the clay if one wishes to extract the contaminants. The next day one will see them on the water as a reddish or yellow oily layer. The water is drawn off by a stopcock as far as convenient. This is done again and the thick liquid again drawn off, then the clay is tipped through a horsehair screen into a shallower box. This is repeated until the good clay in the first box is used up. When the clay has been precipitated from the suspension, the clear water is drawn off then the clay dried either by exposure to moderate heat, or extreme cold, or by mixing it with previously dried and purified clay or with grog, in the proportions given below. This operation, despite its simplicity, attacks equally all the heterogeneities that may be found in clay: light bodies remain on the sieve, pebbles, sand, and pyrites mostly remain at the bottom of the first box, the contaminant clay and the sulphuric acid, if any, are carried away by the water.

---

d. See p. 123 of the *Lithog.* of Pott

Notes D, E
THE NATURE OF CLAY USED IN THE CONSTRUCTION OF MELTING FURNACES AND FOR POT-MAKING

The English now construct their glass making furnaces with bricks made from refractory clay of a dirty white, of Windsor clay, of pipe clay, found near London, and from Stourbridge, &c.

All clays as white as that of Boëlu near Cimey are called porcelain clays but those that have lost part of their *gluten* and which contain more or less detritus, of quartz, principally merit this description; such as those of Sauxillanges and Javogue in the Auvergne and the good quality porcelain clay of Saint-Thiery near Limoges. One would expose oneself to annoying accidents if these were used for glass making without removing the quartz sand that they contain. The reason for this is given below.

Clay is an unsaturated vitriolic salt of earthy base. Many experiments have proved that sulphuric acid is a partial constituent of clay. The sulphurous smell emitted by a furnace in which fresh clay is being calcined is sufficient to convince one.

The base of clay is probably a modified calcareous earth but nothing seems to me to indicate that this base has been calcined. If it were quartz or of the same type as flints, clay would not become charged with sulphuric acid on addition of water and it would not require a strong admixture of fixed alkali salt for its vitrification.

It is however certain, as the celebrated M. de Buffon has confirmed, that clay forms by the decomposition of pebbles, that pebbles are converted into clay. A pebble about four fifths evidently changed into clay that can be seen in the cabinet of the scientist M. Romé-de-Lisle is clear proof. It is no less true that by this metamorphosis the earth of the flints loses its properties and takes on, in part, those of the lime earth from which the flint truly has its origin.

Proof that the base of clay is not completely saturated with sulphuric acid is that clay soaked in water again becomes partly charged and that it then gives alum.

The main properties of clay are to be naturally soft and slippery to the touch; [when] dried in bulk or even fired to a certain extent, to stick to the tongue, which is what the workmen call [in French] *happer*, when appropriately slightly moistened; to have stiffness but also capable of being worked; to decrepitate when large pieces are put on hot coals; not to be appreciably attacked by any acid and to become hard in a violent fire, to the point of giving sparks like flint.

Most mineralogists confuse marl (potter's clay) with the pure clays we have described in our memoir. It is of the greatest importance for the master glass maker to carefully distinguish them; his mistake will cause his ruin. Potter's clay is bluish or greenish, sometimes mixed with red and yellow, becomes red like a roof tile when heated and vitrifies more or less easily in a glass furnace. It owes its natural colour entirely to iron and likewise its artificial red colour and its ease of vitrification.

Pure clays, when carefully selected, do not become red or vitrify in even the most violent heat; they become more or less white except for the purest white ones which become slightly grey for a reason that we will give in a moment.

The colour of grey, brown, and blackish clays is due to a plastic material variable in particle size and abundance. It is easy to convince oneself of this: the colour dissipates on heating or by distillation; these clays giving sulphuric acid more quickly and in greater quantity when they are strongly coloured.

There are two types of white clay, one very different from the other. One has retained nearly all its *gluten*, is very stiff but shrinks a lot on firing. The other has lost nearly all its *gluten*, has virtually no stiffness, and little shrinkage on firing. The latter is mixed with silvery non-ferruginous mica and sometimes up to a third of quartz detritus. The former is never mixed with either mica or crushed quartz. On firing the first loses its whiteness and becomes greyish, the second becomes white. The white clays of Cologne and Boëlu are of the first type and those of Sauxillanges, Javogue and particularly St. Thiriey in Limousin are of the second type.

The *gluten* of all clays is of the same type, a soapy material more or less fine and more or less pure. In the clay of Boëlu, however fine this material may be, it is so abundant that brought together and reduced to carbon by the fire, it is transformed and affects the whiteness.

In the St Thiery clay there is so little of this material that it can escape during firing because of the mica and sand mixed in this material; its carbonaceous matter disappears completely and consequently that, far from losing its whiteness, it can and does become whiter.

The white clay of Bordet near St Germain, Lambron, in the Auvergne is a striking proof of what we have been saying. It is of a striking whiteness, very fine, with neither mica, nor sand nor *gluten*. If mixed with water then fired, far from hardening, it falls to a powder and loses none of its whiteness. It is the best grog that one can use. I have used it with success in compositions for pots.

When calcined more strongly the clay loses only a part of its vitriolic acid.

See our observations on the art of making faience.

* * * * * * * * * *

THE BLENDING OF CLAYS. USE OF GROG

It is not sufficient to have thus purified the clay to use it with confidence, it becomes too compact and shrinks too much during drying. Water rarefied by the action of heat, not finding pores big enough to escape by, would destroy the bond between the parts and would ruin the furnace and the pots. Too much shrinkage would cause them almost as serious damage.

It is necessary to provide the fresh clay with another medium of a nature and in a proportion which will prevent these problems. Where shall we find it? Limestone combined with a clay forms glass under the action of a violent fire. It is almost the same with gypsum. Sand is only useful for parts that are not exposed to contact with glass batch, and in addition it must be very pure, not too coarse, nor too fine. Although crushed glass and iron slag are advised by scientists of the first order, are they not more harmful than useful? I venture to suggest that we shall only find this material in purified clay itself fired to the point where it is no longer susceptible to shrinkage. Clay fired in this way is called *grog*; used pots and the interior of old furnaces provide the best.

If concerned with a new establishment one should make slabs a foot square by 8 to 10 twelfths of an inch thick from pure clay; and after drying subject them for seven or eight days to a violent reverberatory fire, then reduce them to powder very carefully in a mill or better under the pestles of an ore crusher.

The authors and the master glass makers vary greatly in the fineness that they recommend for grog. Pott observed that pots in which fine grog has been used are very subject to cracks and he consequently recommends the use only of large grog from which the fines have been carefully removed.[a] Others have remarked that large grog was the source of voids, of degradations and stones in the glass, and only used the finest. Others (and they are the majority) use a mixture of coarse and fine to avoid these problems; a grog simply passed by a very clear horsehair sieve fails to achieve the intention. One will certainly succeed with pots if one only uses in the mixtures a medium size of grog, that is to say passed by a horsehair sieve, neither too open nor too close, then sieved on a silk screen; the fraction remaining on the latter is that which should be used.

Each part of the melting furnace needs a special grog. This observation is very important indeed. The lower walls below the working holes are the least important part where drops, degradation [of the refractory material], and cracks are of no consequence. It is sufficient to use carefully picked clay and grog passing a very open horsehair sieve.

The crown, or all that above the lower walls, merits great care. Here cracks are very dangerous and attack very serious; partly because fissures favour the formation of tear drops as glassy material collects there then

---

a. See what he says about crucibles in the Memoirs of the Academy of Berlin.

falls into the pots, partly because grains of grog that are detached fall into the glass producing stones which are no less harmful than tear drops. It is advantageous to use only very fine grog in furnace crowns. Degradation does negligible harm to sieges or pot supports but the smallest cracks cause the ruin of crowns, partly because the part sitting above the hearth is always in a bath of glass, partly because the glass batch which spills out of the pots during melting and the glass that falls onto the siege during working find their way into the cracks and make them grow at a rate difficult to believe. It is very difficult to avoid cracks that form in the siege in very considerable numbers. One can only expect to prevent them by using a very porous mixture made only with very coarse grog from which all the fines have been carefully separated.

I know no one who can easily determine the proportion of grog that should be mixed with a given quantity of fresh clay. The proportions given by Pott, in his Memoir already mentioned, seem to me to apply only to very small crucibles which were the subject of his research. In this regard glass works are adrift with dangerous uncertainty. They have no rule for fixing the proportions other than uncertain appearance or the sightless routine of the chief workmen. Nothing, however, is more important. If one is too sparing with grog there is too much shrinkage and cracking; if one is too prodigal there are the problems resulting from defective bonding and strength.

It is commonly thought that each type of pure clay requires a different proportion of grog. That is only true because of the impurities in the clay. If one could purify them and, in particular remove all the sand which is mixed in, I believe that you would find four parts grog to five parts fresh clay would be the best proportions to use.

Can one have a touchstone, an indicator of how much grog to use for each pure clay, without being obliged to separate it from the sand? Make many small cakes four inches square and one inch thick in which grog has been added in varying proportions, without producing too soft or too firm a body, dry them as convenient then tamp once: that cake which only loses 1/18 part of its volume in a violent fire will show you the best mix for furnaces and pots.

The custom in most glass works of lightly firing their clay before mixing it with grog is pernicious: it constantly results in unequal bonding or strength in the products and great uncertainty in the proportions. The clay may be more or less burnt and the different parts never to the same degree.

When the clay, mixed with an appropriate proportion of grog, has been sufficiently trodden and matured, most master glass makers mould it into bricks to use in building their furnaces. Some bake their bricks before use, the others simply dry them well. Both of these procedures seem equally bad to me. It is not possible to make a perfectly bonded furnace interior with baked or dried bricks nor to give it the most advantageous shape: the irregularities are an abundant source of tears and stones and give a very irregular reflection [of the heat]. A violent and continuous fire makes the bricks shrink unequally, the mortar used as a bond flows away and contaminates the glass, the gaps in the joints grow bigger, the heat escapes in a thousand places, the consumption of wood or coal is immense; the quality of the glass less perfect and the furnace has only a short life. How many master glass-makers have not had sad experience of the bad effects of these methods?

Note F
BLENDING OF CLAYS: NEW MATERIAL TO BE USED IN BUILDING FURNACES

Grog is incontestably the best material that can be used in blending clays but one can, for melting furnaces, safely and economically replace it by rolled sand and crushed sandstone. I have myself used crushed sandstone with good success. Quartz sand which has not been milled and has retained its corners, has not succeeded so well. Furnaces in which blends containing quartz sand have been used have constantly sagged in the most exposed parts such as the tops of the arches over the hearths or the upper part of the crown, &c. at the highest degree of heat.

This phenomenon seemed to me too unusual and too interesting not to seek the cause. I found that, although surrounded by clay, the quartz grains have lost their corners and become rounded and that the grains in contact with each other are stuck together; this incipient melting and loss of the corners has decreased the volume of the quartz grains and caused considerable shrinkage which has necessarily upset the centre of gravity of those parts of the furnace of which we have spoken.

This alteration of grains of quartz sand seems to me to prove that quartz is more fusible than ground sandstone or milled sand and in consequence it would be dangerous to substitute the former for the latter in blends of clay.

In districts where quartz is very common, in regions of ancient creation, such as the Auvergne, Limousin, &c., all the sands are of quartz. It is thus important not to allow them to be mixed into blends to be used for melting furnaces.

There is a simple way of telling whether a sand can safely be substituted for grog; that is to expose it for twenty four hours to the greatest heat of a German type of furnace or a bottle-glass furnace. If, after this test, the sharp corners have not disappeared nor the grains become stuck together, it can be used as a substitute for grog.

One must not conceal the fact that melting furnaces composed of fresh clay and grog are very expensive, especially for master glass makers who have to bring their clay from a distance. These furnaces demand a considerable amount of labour and require the minutest attention to very many details. If one omits any of this attention one runs large or small risks which may be correspondingly dangerous.

Is there no more expeditious way of building glass making furnaces? Is it impossible to find a material less expensive than pure clay which would demand less labour and require less attention, which would suffer fewer inconveniences, be exposed to fewer dangers, assure greater durability and a longer life?

Slag from iron works, which until the present has found no use in

France, offers us this material which combines such great and such precious advantages.

But can one construct from glass, you will say, a furnace intended to found and make glass products? This proposition will seem incredible to many people; nevertheless it is not only possible but also as easy as it is sound. It is easy to cast furnace slag into a mould that opens which, placed on a good slab of cast iron and having a suitable shape, allows one to cool the slab of glass very slowly in a front corner of the furnace so that it remains in a single piece. A furnace may then be built on a circle from these bricks, filling the joints with a grouting made from the same slag ground and sieved very finely. The whole crown is then covered with a layer of two or three inches of ordinary clay of the consistency needed to be moulded into bricks. In this state the inside of the furnace is simply held at cherry red heat for ten days and then, without any fear, raised to the greatest possible heat. At this stage the furnace material is no longer glass. The bricks and the grouting have fused together forming one solid body and the slag has been converted into porcelain of the most refractory kind which is the most capable of continuously resisting the action of the greatest heat that I know.

This singular and useful metamorphosis is solely due to the property possessed by the slag of converting itself, without cement, into porcelain at a moderate heat. All crude glass, such as black bottle glass, has the same property. It is easy to convince oneself even in a kitchen fire. If we have recommended that the crown be covered with clay bricks, it is so that the exterior of this part of the furnace can also be raised to red heat and converted to porcelain.

To dissipate all doubt and to reassure timid and ill-informed spirits, I will add that in Sweden they construct all the upper parts of their blast furnaces, above the *tuyères*, from slag bricks and that they find this procedure very satisfactory; one could use these bricks for the whole height of the lining in these furnaces.

Note G
DISADVANTAGES OF COARSE GROG

A director of plate glass manufacture at Saint-Gobain having blindly followed the advice of Pott, caused very considerable losses to the company. Furnaces built from, and crucibles fabricated with, blends which contained only coarse grog were attacked much more quickly, leading to much less satisfactory performance, much shorter life and drops, cords and stones affected much more of the glass. In large scale manufacture it is always very costly to recover the proper method of working and return to good production. Saint-Gobain has had the sad experience. Nothing is more unfortunate for a factory than to have no other guide but blind routine.

Note H
BAD EFFECTS OF TEAR DROPS IN FURNACES

It is important to avoid drops, even for common glasses such as ordinary window glass and bottle glass; they are the source of very disagreeable cord, the glass from which they are formed being always less homogeneous and having a different thermal expansion, it is rare that they do not cause breakage of the glass during annealing or in the warehouse.

* * * * * * * *

BUILDING THE FURNACE

A means of avoiding all these problems [excessive shrinkage of clay and cracking of the furnace structure followed by rapid corrosion] is known in the large glass works of Normandy and in some of those in the Palatinate and is very easy to put into practice even though Blancourt assures us that the secret is kept within the family of one master mason.

The prepared clay mixture is formed into four types of tile: some a foot square and two inches thick for the lower walls and part of the crown, tiles twenty inches long by six wide and two inches thick for the arches, tiles a foot long, ten inches wide at one end and seven at the other and two inches thick for the crown, and tiles for the siege which are two inches thick and of a size sufficient that, when laid, the sieges are only six inches apart at the bottom and two fifths of the furnace width apart at the top but only twenty to twenty four inches high according to the size of the furnace. If these tiles are made thinner, as is commonly done, the joints are multiplied unnecessarily; if they are made thicker they do not sit perfectly one on the other. When these tiles have hardened somewhat they are lightly brushed and tamped with a wooden bat. They are used in this state to build the furnace.

It is important that the main body of the furnace where the vault opens between the sieges and where the four open galleries or arcades meet at right angles at the other end (if one wishes to use coal) be extremely solid, that the hearth be able to resist the greatest heat (hard gritstone is excellent) and that the four arches be made with the greatest care. See figs. 1 and 2.

The square of the furnace being drawn out and made from white bricks or scrap clay and sand, and the arches being built from good ordinary bricks, it is convenient to begin construction with the sieges. It is then much easier to lay the tiles and set them tightly together by vigorously beating them than when the sidewalls and the crown are already built and there is no risk of causing dangerous shocks: everything is very solid. Then one makes a bed for the sidewalls from the tiles; one wets this bed with a broom using a clean grouting made with the tile mixture. This is then covered with a second layer which is well tamped down and so on until the furnace is complete.

The arches for the working holes and the small flues are built using pieces of wood of the shape desired for these openings.

The outside and the inside being matched up and bonded together, the furnace should be dried very slowly, tamping it many times every day. If cracks form during tempering or drying out of the furnace either because the mixture used for the tiles was a little too plastic, or because it was heated too rapidly, or because the space between the furnace and the

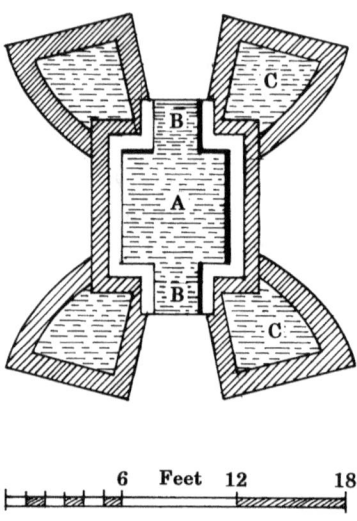

Figure 1. Plan of a wood fired melting furnace
A Hearth of the furnace; BB Stoke holes; CC Annealing arches

arches was not filled completely, there is no need for alarm; the remedy is very simple: the fire is carefully damped down then, when the furnace is cool enough to enter, the crevices are caulked with hemp well filled with the mixture used for the tiles which is driven home as well as possible with a strip of iron about an inch wide and 15 to 18 in long. The furnace is then heated by degrees.

It is easy to see that by this method there is never: 1, any fear of unevenness nor irregularity of reflexion [of heat]; 2, the inside of the furnace can be given the most advantageous shape without harm to its strength; 3, that a furnace constructed in this way, having no joints, at least obvious ones, does not lose anything in the fire and can last ten to twelve years if one is careful to repair when necessary the hearth, the sieges, and the working holes and with it to see in many excellent New Years. These repairs do not involve any difficulty. The only care that one needs to take is not to allow the furnace to cool too rapidly. This will avoid spalling.

There are at least as many other ways of building the furnace as the glass industry has branches. In making bottles or window glass (the crown process or another) the furnace has six arches, one at each corner and one above each stoke hole: in most of those used for goblets, drinking glasses, carafes, crystal, &c. there is only one arch above the furnace: in others, especially flat glass, Bohemian style, there are four, one at each corner of the furnace. Nothing is fixed about the shape or the capacity of these arches, the position, the shape, or the diameter of the opening through which they receive heat from the furnace. We are no further advanced concerning the shape of the furnace interior. Only the chalk line of the master mason, or the founder, regulates what should be fixed by deep understanding of furnace technology. Each has his own dimensions. Some prefer a round shape and a well rounded crown and place in their furnace three, five, seven or eight, or even nine pots. Others like their furnace to be square up to the height of the middle of the working holes and that the crown be more or less flattened; some make a type of truncated quadrangular pyramid: I know those who, to avoid drops, make the crown very flat on the sides over the hearths but as steep as possible on the sides with the working holes. Almost all are obliged to knock down part of the furnace to put the pots in place and rebuild it afterwards. All of these furnaces heat feebly and consume a very great quantity of fuel. It would be very tedious and perhaps not very useful to the art of glass making to recall the faults peculiar to each type.

It is as difficult as it is important to find a furnace suitable for all the operations of glass making; such a furnace ought to have six or ten working holes, four small flues [going to arches], two stokeholes, and produce the greatest possible heat with the least fuel. It is for the Academy to decide whether I have come close to the solution of this problem. I can give assurance that the furnace of which I give the plan and section in figs. 1, 2, and 3 refines glass very well and quickly with a third less wood than those that are less perfect [the text says *imperfect*]. From five to eight feet are the most advantageous dimensions. Greater or smaller and the glass does not refine so well and proportionally more fuel is used. Six workmen can work this furnace at the same time and anneal their production very well in the arches by using clay or sheet iron trays, containing a sprin-

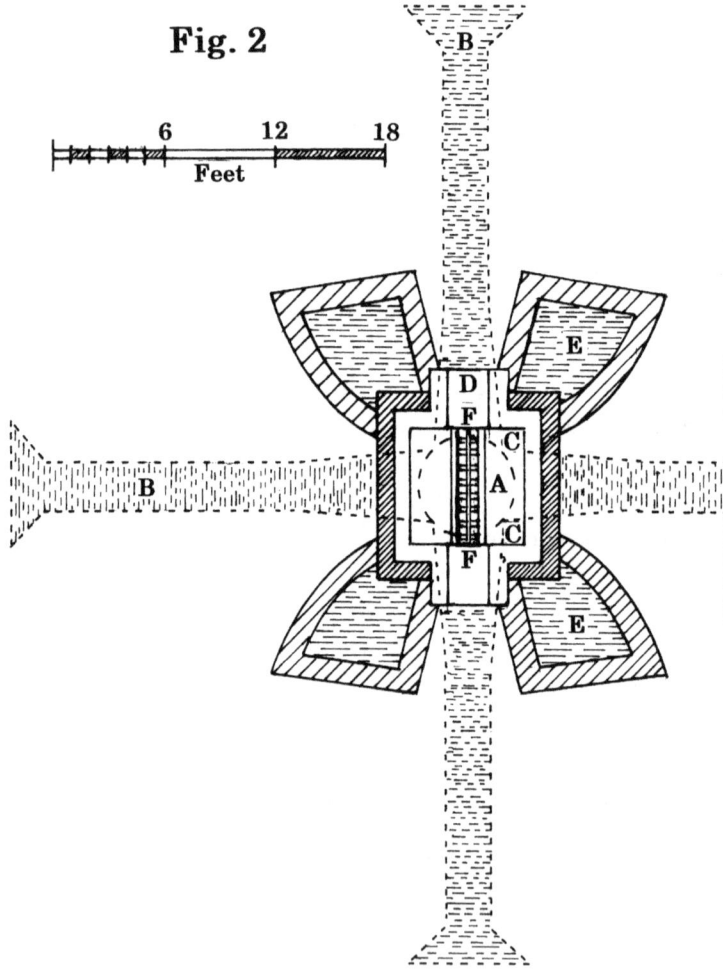

Figure 2. Plan of a coal fired melting furnace
AA: Hearth of the furnace; BB: Air flues; CC: Sieges for the pots;
DD: Stoke holes; EE: Arches; FF: Grate.

On Glass Making

Fig. 3

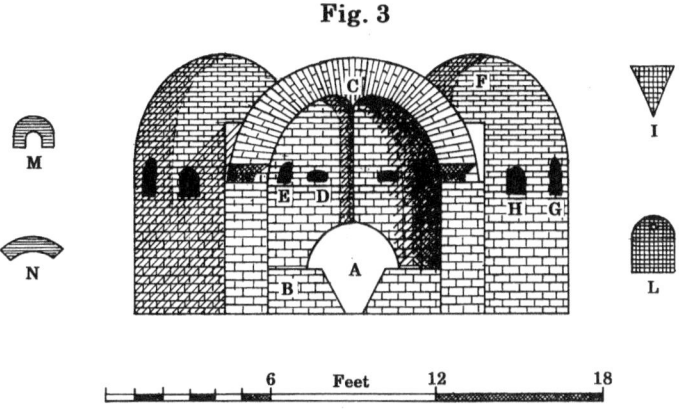

Figure 3. Vertical section through melting furnace

A: Arch of stoke hole; B: Sieges;

C: Rib dividing the interior of the furnace into two equal parts across its breadth and reflecting the greatest possible quantity of heat onto the crucibles;

DD: Small flues to the arches for materials or for fritting the batch for bottles or sheet glass;

EE: Flues to annealing arches; FF: Arches for fritting or annealing;

GG: Holes for putting in or removing pots from the annealing arches;

HH: Openings for placing in the arches bottles and trays or pots for annealing;

I: Stopper for working hole during melting and refining;

L: Stopper for working hole during standing off;

M & N: [Not described: possibly pieces to restrict stokehole when required].

kling of clay, a foot in diameter by two deep and closed by a stopper in the Bohemian fashion. If one wishes to blow lamp glasses in the English style or window glass it is convenient to reduce the thickness of [the walls adjacent to] the working holes to two thirds on the inside of the furnace, Those who wish to make their frit with the heat of the melting furnace can conveniently do so. They have only to make a vault beneath each of the annealing arches with an eye at the height of the siege. One could cast large plate glass for mirrors by making four working holes nineteen inches by seventeen wide, one at each end of the two sieges.

Note I
ENGLISH COAL-FIRED FURNACES

In England the coal-fired furnaces for making glass for chandeliers, crystal and white flint are only slightly different from our old French furnaces but much more solidly built. The hearth above the grate is at the centre of the furnace and appeared to me to be about two feet square. The coal is pushed there through an aperture made between two pots and at the level of the siege of the furnace on which the pots are placed. This hole is never closed except with coal itself.

The smoke and the ash of the coal cannot affect the colour of the glass because the flames do not act directly upon the material being melted. This is to say that the pots are covered on the side next to the fire and only open on the side facing the working holes [where the openings are] of the same size as the working holes. These English furnaces are heated very briskly and appear to me to be economical. The apertures by which the flames escape upwards, above the crown, are made between each pair of the pots.

* * * * * * * * * *

CONCERNING POTS

There are two ways of making pots; that is *by hand* or in *moulds*. The latter is restricted to small pots, the former to large ones. I can give assurance, after a sufficiently long experience, that it is easy to make large pots in moulds and to do so with advantage.

1. It takes a long time to train a good hand pot-maker: it is scarcely necessary to guide a workman a couple of times for him to be able to

make good pots in moulds.

2. It is rare that a pot made by hand is raised true: it is always so in moulds.

3. One can only beat gently a hand made pot; one can do so as much as one likes with those made in moulds.

4. The most skilful pot-maker can make by hand only four pots of about 30 in. each in a week; a workman can easily make three a day in moulds.

There is no shape which has not been given to pots. They have been seen square, triangular, with many sides, as broad at the base as at the top, pot-bellied, with the diameter in the middle greater than either above or below, and more or less widened at the mouth. The three first types suffer from unequal expansion because of their unequal thickness, failing readily at the corners and they also are less use for the *purifying* of glass than the following two types. Glass melts and refines much quicker, other things being equal, when it offers the greatest surface to the reflected fire. According to this very true principle it seems to me that the pots should be reversed cones but not carried to a point because they would lack stability and one would waste too much material. There is a proper mean: this is to make the bottom diameter 1/7 less than the top diameter. This taper, far from impairing stability, contributes to it; it is very suitable for the purifying of glass and allows the pots to be proportioned to the capacity of the furnace, for which purpose they must neither be too big nor too small.

The best crucibles are those that have been made with pure clay prepared and blended as has been described above. They should be treated with the greatest care; of a medium thickness, an inch and a half in the shoulder and two inches in the body and the base, of an appropriate form, tamped with care, dried very slowly, and fired with a violent and long-continued fire.

Note L
THE ENGLISH SHAPE OF POT

Concerning the care needed in the manufacture of pots, one may refer to our memoir on the perfecting of the small Auvergne crucibles.

Regarding English pots, it is indispensably necessary that the hood should be fully arched. If it is a little flattened it will bend and restrict both refining and working. With the normal height of the crucible one makes, in the English type, a mouth rather larger than that used for the working hole. When rebuilding the wall demolished to set the pot in the furnace, using clay of the same type as the pot, one reduces the opening, as required to prevent entry of the flame into the pot. The edges of the pot mouth are attached to and built into the interior of the front wall.

The hood of the pot should only be made by hand.

* * * * * * * * * *

CONCERNING THE MATERIALS TO BE CONVERTED INTO GLASS

None of the four types of clay or stones that we know, *calcareous, gypsum-like, clayey* and *flint-like*, can be vitrified by themselves even in the most violent heat.

The experiments of Pott[a] do not allow us to doubt it. To be changed into glass each requires the addition of something else, either metallic bases or fixed alkali salt. There is no other which forms, with a given proportion of the latter, so clear and transparent a glass as pebbles [of silica]. This is also the combination best known in the glass works and, perhaps, the only one that it is necessary to know.

The type of stone known as rock crystal, quartz, flint, sand, sandstone, &c. is the principal constituent of glass. These stones, except for rock crystal and quartz, are very common in France. Very good white and

---

[a]. See Chap. IV of his Chemical Examination of Stones

On Glass Making 61

semi-transparent pebbles may be found on the sea shore, along many rivers, &c. Black, brown, off-white, opaque, and coarse [silica] pebbles are assuredly not rare but it is sufficient to heat them to redness and quench them in water to make them perfectly white. We have a large number of deposits of very good sand and, what is perhaps more valuable, sandstone in abundance. There are very few glass works which use the latter in glass batches, Merrett[b] even says that *it cannot be used for this purpose,* without doubt for fear that it would cost too much to crush it, but one has only to heat it to redness in the arches whilst making the melts, quench it in water, and then beat it lightly. The operation will certainly not be costly unless one has a rare and hard type of sandstone.

Glass made with paving stones or others of a whitish or yellowish colour is no different in hardness from that made with sand, although *Wallerius*[c] states that.

The most highly coloured and clay-rich sand is the best for bottles, because it requires little alkali and makes a very stable glass. White sand or sandstone may be used for green glass and even for white flint. The red or yellow streaks that one finds in it usually disappear on heating. However, if one does not have good flint pebbles, how can white sand or sandstone be purified sufficiently to make the best and finest crystal? Repeated calcining and quenching is often insufficient. The iron earth from which they are rarely free, and which always colours the glass to some extent, resists that. The celebrated Stahl[d] recommends washing them with care in river water containing a little nitric acid but will this method not be very costly and can one expect it to achieve its objective unless the sand or sandstone is reduced to very fine powder? Aqua regia and Glauber's proserpine (butter of antimony), with which one can certainly achieve the desired extraction, are too difficult to prepare and too expensive for a

---

b. See the notes on the first book of Neri
c. See p.141 of the first vol. of his *Mineralogy*
d. Stahl *Fundam. Chym.* pact. 2, sect III, cap. III. *de modo conficiendi varias gemmas artif.*

master glass maker to decide to use them. I shall demonstrate a method just as effective, much simpler and less expensive. One has only to mix, by dissolution, a hundred pounds of sand or ground sandstone with four pounds of glass gall, put this mixture in an old pot, or in all of the pots at the end of a founding and subject it for seven or eight hours to the most intense heat: the gall disappears, dissipating every atom of colouring matter, and the sand remains white as snow, very pure, suitable for the best crystal and even to make imitation precious stones. Many authors have suggested that there are different types of sand or flints which melt more easily than others. This is only true on account of the impurities (especially ferruginous matter) with which they are contaminated. It is wrong to believe, like Merrett[a], that crystal requires a soft sand and common glass a hard sand.

Note M
FUSIBILITY OF DIFFERENT TYPES OF SAND. CONJECTURES ON THE ORIGIN OF QUARTZ

At the time when I wrote that all pure flints were equally easy to melt I was not sufficiently acquainted with semi-transparent milky quartz. I had not had any occasion to use it on a large scale, the true means of knowing the properties of minerals and avoiding misfortunes in the useful arts.

Quartz in the mass becomes vitrified over its whole surface and is covered by a skin of glass, when simply exposed for twenty four hours at the ends of the small flues which heat the arches adjoining the melting furnace. Vitrification occurs more quickly and more deeply if a fragment of quartz is put into the furnace between two crucibles. In glass batches in which one uses quartz, one twentieth more quartz should be used than of any other type of sand.

At Cleuzel, near Langeac in the Auvergne, considerable blocks of

---

a. See *The Art of Glass*, French translation in quarto, p.17.

quartz are found which are milky in some parts and transparent in others. The latter is as transparent and colourless as the massive rock crystal from Madagascar. Different colours are also found there but the colours are not very true. I have also found there smoky quartz and in all types I have noted threads of black schorl [tourmaline].

This transparent quartz seems to me rather less fusible than the milky but easier to melt than the rock crystal in needles, which seems to establish a difference between these two types of rock crystal.

I have not had the opportunity of testing massive rock crystal from Madagascar but I have every reason to believe that it is as easy to melt as that of Cleuzel.

Is transparent quartz a product of fire? This problem is more difficult to resolve than it appeared to the greatest writer of our century who gave an affirmative answer. Without showing a lack of the respect that we give to so imposing an authority we must state a few doubts.

If this massive crystal were the product of fire, how can one conceive that one part of the block should remain milky although subjected to the same degree of heat? It is no less difficult to imagine that the black schorl, itself easily melted, should be found regularly distributed in the transparent quartz which is much less easily melted and of very different density. In addition, when one adds black schorl to a glass batch it gives the material fluidity and colour. After cooling very slowly in the pot no vestige is visible.

Achard has demonstrated by a decisive experiment, which has been repeated by Magellan, that crystallization of rock crystal is due to fixed air, to atmospheric acid and the celebrated Bergman has made rock crystals with flints and the acid of phosphoric spar. It is not possible, by simple calcination without mixing together with other materials, to make pulverized quartz as fusible as these other sands but this same means also removes from the latter a part of their nature. What use we shall make of this observation will be seen later.

M. Sage of the Academy of Sciences, &c., considers quartz and all rocks of the same type as flint to be a natural sulphated tartrate. What he says to support this opinion, see pp. 242–4 of the first volume of the *Elements of Mineralogy*, does not seem convincing to me. It is only certain that the batches to which he added lean raw potash or badly purified fixed alkali mineral salt, materials containing the colouring principle, melted with *considerable effervescence* and *foaming*.

Nothing similar is seen during the softening and melting of batches made with well purified sand or fixed alkali, either mineral or artificial, purified with great care by dissolution and calcination.

I have never seen gall on glass during melting when fat potash, soda ash or recrystallized soda containing Glauber's salt and sea salt have been used in the batch. I venture to say that there is never any potash, which does not contain a larger or smaller proportion of sulphated tartar. I have used such potash which contained almost a third of this neutral salt; we shall see in what follows that it is possible for all the fixed alkali in ordinary ashes to be converted into sulphated tartar, even though one has not used for its extraction water containing selenite [gypsum]. On many occasions, this neutral salt being, in my opinion, undesirable, I have extracted considerable quantities of it from potash by a very simple method.

Having dissolved the potash in the minimum possible quantity of hot water I tipped it onto a woollen blanket folded in four; the fixed alkali passed through in the water but the majority of the sulphated tartar remained on this coarse filter. In this way one could, if one wishes, produce large quantities of very pure sulphated tartar given the availability of sufficient rich potash.

It is no less certain that there is no soda which does not contain some greater or smaller proportion of Glauber's salt and sea salt. All chemists can easily confirm this by experiment.

Thus the origin that M. Sage gives for gall seems to me more than uncertain; the observations above prove this. One cannot conceive moreover the reason why, during the melting of the materials, the acid of natural sulphated tartar gives up its base to combine with another fixed alkali of the same nature. Given that the composition of quartz is not perfectly understood, it seems to me that no one is better placed to make, with success, the researches that this topic merits than Sage.

The beautiful discoveries of Achard and Bergman seem to establish that quartz is composed of an earth analogous to that of alum and atmospheric acid.

* * * * * * * * * *

FLUXES

The second principal material used in glass batches is the flux; fixed alkali salts, mineral or artificial, or a mixture of fixed alkali salts together with their ash-like component, an alkaline earth of saline nature[a]. There is probably no part of the art of glass making over which the authors and master glass-makers are less in agreement. Each has his favourite flux.[b]

Agricola says that in his time *nitre* was preferred. *Merrett* states that *time and experience had led to the abandonment of nitre* as being *too soft and weak, producing gall, and that Rochetta or Syrian powder had taken the first place*[c], *Kunckel* and *Henckel* state, following *Merrett*, that *glass made with soda is not esteemed; that it breaks very easily on cooling, that it always has a bluish tint*; that, *in a word soda, even when mixed with manganese, never produces a good glass*. One can easily see that these authors lean in favour of *potash*.

The first even says that the crystal that he has made *is superior to* those that Neri had made with great trouble and in which he had used fixed alkali salt of soda, Rochetta, alkali tartrate, nitre, &c.[d] *Gellert* thinks that soda is better than any other salt extracted from ash for making good durable glass; that glass made with potash is more subject to attack by acids than other[e] kinds and even that it decomposes in air. At the present time

---

a. See p.175 of M. Pott's *Lithog.* and p.94 of the continuation
b. See Book 12 of *De Re Metall.*
c. See p.29 in *The Art of Glass* (quarto translation).
d. See *Op. cit.* pp.11, 103, 561.
e. See pp. 25 and 26 of the French translation of his excellent *Chemical Metallurgy*.

the glass houses in Germany use only potash. In those of France and Italy current usage is largely restricted to soda and saltpetre. To which of these fluxes, rejected by some, adopted by others, should we give preference? To that which is most easily obtained and involves the least expense. I can state, after very varied and frequently repeated experiments that all are equally good, provided that the preparation is appropriate, the proportions in the batch just, and the melting and purification sufficient.

Note N
FALSE IDEAS OF AUTHORS ON THE CHOICE OF FLUXES

The contradictions on the choice of flux are obvious proof that the authors we have cited in our memoir had only a superficial knowledge of the art of glass making and that they followed the operations and observed the phenomena only imperfectly. It is clear that these authors and master glassmakers have decided in favour of one flux rather than another only for reasons foreign to the nature of these salts and to the art of glass making.

Agricola only made nitrate his favoured flux only because in his time this salt was the best known and least expensive. Merrett only rejected this and preferred *Rochetta* because the purification of saltpetre, especially the way of extracting sodium chloride, was little known and that commerce between Europeans and the Levant had made Syrian powder very common. The manufacture of potash began in Germany. It is therefore not astonishing that Kunckel and Henckel made it their preferred flux. I am not aware that soda was in use in the glass houses of Saxony. Consequently I find it impossible to divine the reason why Gellert decided to use it exclusively. I suggest that none of these decisions, however respected the name of the author, was made on a sound foundation.

* * * * * * * * * *

Fluxes can be found in France in at least as great abundance as in foreign countries. However rash this assertion may appear to be, it is nevertheless well founded. We can make annually a very considerable

On Glass Making

quantity of red or white potash.

[*The* Collected works *here includes a long footnote extending over three pages which is almost identical to parts of Note O but the latter is more extensive. The footnote is therefore omitted here.*]

There are many provinces where immense quantities of small wood are lost, having been ignored, and where wood cutters neglect the ashes that they themselves make in the course of exploiting a tract. Bracken is very common throughout the kingdom: the ashes of this plant, cut towards the end of July and burned in the same way as soda, gives about one ninth fixed alkali salt. The same may be said of grape marc: its well made ashes produce at least as much as bracken ashes. The ashes of what remains in the stills of brandy producers have given me up to one fifth of very good fixed alkali salt: it is rare that these distillers do not mix liquid lees with the marc. From the little that I know of our wine production, it is conceivable that we could obtain annually from marc many millions of pounds weight of very good *potash*. It is surprising that such a useful and easily obtained object has been neglected until now. The ash of tobacco offers us a new resource. So long as it is made with care, it produces more than one third of its weight of fixed alkali. It is clear that we could very easily procure for ourselves, and without foreign help, the potash that our glass houses, bleachers and crude soap boilers need to consume. Without doubt this branch of industry deserves to be encouraged. The master glass-makers are those most interested in this and it would be very easy for them: they have only to stimulate the peasants to make trials, to give them certainty of sale and give them direction. We will later give them the means.

Note O
RED AND WHITE, LEAN AND FAT POTASH. ADULTERATION OF POTASH; MEANS OF DISCOVERING IT

Red potash is the fixed alkali salt extracted by leaching and evaporation from the ashes of all plants except maritime plants. It contains a great deal of yellowish colouring matter.

White potash is only the red calcined by a reverberatory fire to relieve it of its yellow colour, its yellow colouring matter and the majority of its atmospheric acid.

Large quantities of potash are made in Alsace, Lorraine, and in the Ardennes. This manufacture is beginning to extend to other provinces. In the main towns there are people who buy it pound by pound from the peasants and wood cutters. Cupidity has led to many fraudulent practices in this branch of commerce. It is important to make known the most serious.

One of the most dangerous for makers of *fine* glass is the mixing of sea salt with the alkali or selling as potash the salt extracted from the ashes made under the boilers of salt works. There are three equally sure methods of detecting this fraud: this potash melts easily in the heat used for calcining, takes on only a pale blue colour, and produces glass from only a fraction of the fixed alkali that it contains. I have received samples of it containing more than half sodium chloride coming from the salt works at Dieuze. This fact seems to me to merit the attention of the minister. By correcting this abuse he would render an important service to the art of glass making.

This potash, containing a large amount of sea salt, can be very useful in works making bottles and ordinary glass only to help the materials to melt and to remove from the glass the excess of colouring matter but it is a matter of justice and good faith to sell it only for what it is and not to sell it as good potash.

It is known that old ashes provide a greater quantity of alkali than new ones and this is a certain fact. As a consequence they are left a long time dampened to *germinate*, exposed to the open air but sheltered from the rain, or kept in slightly humid places where the outside air has free access. These ashes give more alkali for the reason that a part, often considerable, of the fixed alkali is converted to sulphated potash*. One can see more about this in my memoirs on the cause of bubbles and smears in glass and on the manufacture and commerce of potash.

There are potash manufacturers who mix soot with this fixed alkali salt and this gives the glass a very marked yellow colour. Others, and these are the majority, do not allow their lye to clarify. This results in the potash being weaker and contaminated with a greater or lesser quantity of earth and that it is very difficult to purify, of the colouring contaminant by calcination, especially when the ash is made entirely or partly with ashes of oak[a]. See my memoir on the *Manufacture and commerce of potash*.

One can divide red and white potash into *lean* and *fat* potash. They are so named because the latter usually makes the glass cloudy or milky. The former contains very little or none of the sulphated tartar which makes the glass bubble, froth and even foam in the pots and makes it very difficult to eliminate the yellow colouring matter. Fat potash contains a much greater quantity of sulphated tartar than is provided by ashes aged for a very long time. In melting it gives a lot of very white gall and a part of the neutral salt, not finding in the batch sufficient fatty or colouring matter to be made volatile, remains as a more or less dense mist in the glass and infects it with specks, clouds and milkiness. It is much rarer to see such defects in glass using mineral soda as alkali than in that made with potash because Glauber's salt and sea salt, necessarily finding a greater quantity of colouring matter in the mineral alkali than the sulphated tartar finds in the artificial alkali, are much more volatile. See my observations on the art of making faience.

---

* *The omitted footnote O here adds* "a salt harmful to glass making when large proportions are found in the potash".

a. The omitted footnote O here says 'pine' not oak.

To avoid these inconveniences of lean or fat potash and achieve equally good production it is very useful to mix very well a large quantity of potash of one type, either red or white, in a very large tub; [this implies mixing the contents every time a small addition is made]. This is the surest way of avoiding the pernicious effects of either lean or fat potash. It is easy to see that an excess of either too lean or too fat will be corrected by the other and that, over a long period, the quality of the flux used will remain of the same. It is important that this tub be carefully closed to exclude dust and humidity as soon as the alkali needed has been withdrawn each day.

* * * * * * * * * *

SODA

Without speaking of marine algae known under the names of kelp and seaweed, which are very common along our coasts, we have another much more precious resource, namely *soda*. It is made in Provence and in Languedoc and, simply fritted with sand, is proper for making good glass. Experiments that have been made in Poitou are very hopeful. Many good types of kali, especially *kali majus cochleato semine* grow naturally on the coasts of Provence, Languedoc, and Roussillon[a]. I have a very good reason to believe that *kali of Sicily* and the most esteemed of the Spanish types *capillary-haired kali, broom-leaved kali and tamarisk-leaved kali* could be grown there with success. It is more than probable that by appropriate culture and incineration of these plants, we could procure for ourselves very abundantly and in places incapable of producing any other useful thing, soda equivalent to the best from Spain. That which is made on the coast of Narbonne and in the so-called *Isles les Saintes*, only from *kali majus cochleato semine*, does not yield much to them in quality. These sodas give as much fixed alkali, about half their weight, and they show no other differences except for slightly more sea salt and Glauber's salt.

The soda or *salicor* that is made at *Pérols*, a new town near Montpellier,

---

a. See p. 536 of the *Natural History of the plants of Europe* is much inferior to that from the *Isles Saintes*. It gives scarcely one third of fixed alkali and contains a much greater quantity of neutral salts; no doubt because they burn together with the *kali majus cochleato semine, kali majus geniculatum*, maritime *absinthe* and *fennel*, as well as many other maritime plants. If only we cultivated, in our southern provinces, *kali majus cochleato semine*, its ash would give less fixed alkali but we could make our soda equivalent to the best from Alicante or even better. The means is simple. We have only to wet the ashes of *kali majus cochleato semine*, still red hot in the pit, with a strong lye of the same ash.

Although the States of Languedoc, always attentive of the good of the province, have decided to encourage the culture of soda plants; although this culture and means of improvement that I have proposed would be as easy as they would be lucrative, we may perhaps have to wait a long time for soda of a better quality. In the meantime, whilst waiting, it would be very advantageous to substitute our sodas, such as they are, for foreign ones; a single operation would put us in a state to do this. It is only necessary to extract the fixed alkali salt and to use this instead of the natural soda. I can assure you with confidence that this salt, according to its degree of purification, is proper to use, to make any type of glass from common white up to the most beautiful crystal. The earthy part of all sodas, like the ash of all vegetable matter, is of exactly the same *alkaline* nature; the colouring matter mixed with it is precisely the same in all, *phlogiston*. The fixed alkali salt extracted from the ash of all vegetables, without exception, is the same in relation to glass making where it is equally proper to make glass of a good and beautiful quality. It is easy to confirm the truth of these three propositions which seem incontestable to me. Whence then does the difference between our sodas and those of Alicante arise?

This can only be that either they contain a larger quantity of earthy constituents, or that their fixed alkali contains a greater proportion of neutral salts: but there is not one twelfth of these neutral salts in the salt obtained from the worst sodas of Languedoc. One finds as much in most potashes. This quantity, provided that the batch is well made, does no harm at all to the quality of the glass. It is possible to make glass very well and very beautifully with the fixed alkali from *kelp* which contains more than half its weight of sea salt and Glauber's salt. It is thus solely in the greater proportion of earth that one should seek the cause of the essential difference between our sodas and those of Alicante.

Batch made from equal parts of good sand and the better Spanish soda, in which the fixed alkali salt is present in almost the same proportion as the earth, *frits* well, bleaches easily in a reverberatory heat, melts without trouble, and produces a passably transparent glass. Equal parts of good sand and ordinary Languedoc soda *frits* very badly, always remaining a yellow brown, and does not melt in the most violent heat. If one reduces the proportion of sand it makes a glass of poor transparency and very disagreeable yellow-green colour. The manganese* [added as decolorizer] can only, in this case, make the glass less transparent and the colour more unbearable.

Why does the latter give a product so different from the former? It surely can only be because it contains about two parts of earth to one of fixed alkali: separate the latter from the earth, from the ashes, and the differences disappear and the difficulties are removed.

But how, some of the master glass-makers may say, can we extract the soda salt without increasing our expenses? This would be difficult if I proposed to them the method of *Neri* or that of Kunckel[a].

I will tell them a much simpler and less expensive method (see Fig. 4). One puts the pulverized soda, one part to eight parts of water, in the boiler D1, three quarters full of water. The water is hot enough to dissolve the salt easily and become charged with it, but not boiling, so as not to hinder prompt precipitation of the earth. When it is well clarified it is drawn off by a stop cock into the middle boiler D2 under which there is a large fire and from there, to avoid retarding the evaporation by too great a

---

* [The original has *magnesia*, then a common confusion.]

a. See pp.2, 11 & 307 of the *Art of Glass*, French quarto translation.

On Glass Making

Fig. 4

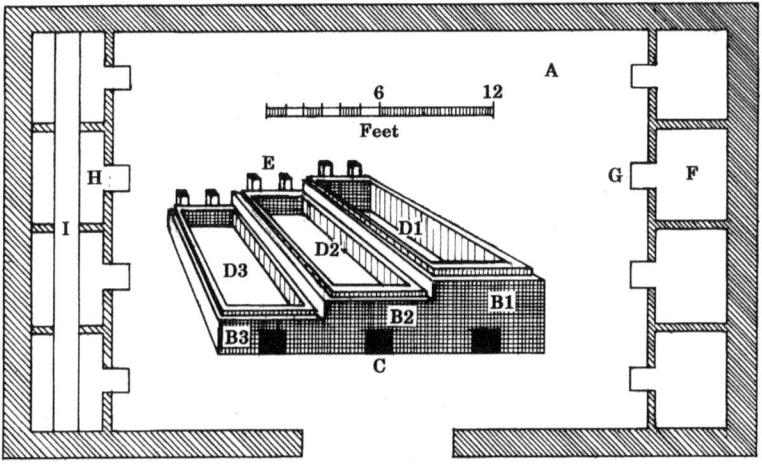

Figure 4. Plan of a workshop for extracting fixed alkali from soda
A: Interior of the shop (plan); B_1, B_2, B_3: Perspective view of boiler for extraction; CC: Stoke holes; DDD: Boilers seen in perspective; EE: Chimneys; FF: Storage bays for finely powdered soda and water for dissolving the salt; GG: Reservoirs into which the lye is put and from which it is taken to the hottest boiler D1; HH: Vats into which the salt drains after being placed in the trough I.

thickening of the lye, into the third boiler D3 where the salt is precipitated by a gentle heat and from which it is taken with a type of iron skimmer to let it drain on an inclined sheet metal tray above the same boiler. The two walls separating the furnaces should be only eight inches thick and each of the two holes [in the walls?] two inches square in section. The fire is only made in the chamber in the middle, the two others are heated only by the heat which it communicates to them and by the embers that it provides. The earth from the soda is carried into the compartments F,F: it is wetted many times with water to extract all the residual fixed alkali that it may contain and this water is taken in small basins to the boiler D1. The extraction of the salt only costs three pence a pound. Four men with the equivalent of a cord of charcoal can extract a thousand pounds in twenty four hours.

Nothing is more expeditious to pulverize the soda than a *crusher* operated by either water or wind. This machine is so well known that I do not think it necessary to give a description[a]. Crystal is much finer if the fixed alkali has been well purified. But of what does this purification consist? Following the authors on the art of glass-making and a great number of chemists, it consists in dissolving the fixed alkali and drying it, by evaporation, many times. There are those who prescribe such dissolution, filtering the solution and evaporation up to five times[b]. This method is tedious, very expensive and insufficient. It is costly not only in time, the instruments, and the wood that it requires but principally because of the loss of fixed alkali salt that it occasions: not two thirds remains, and it contains a greater proportion of neutral salts. A single dissolution is sufficient to achieve a good enough separation of the saline part from the earthy part. The earth that one obtains in the different dissolutions is less proof of the insufficiency of the first dissolution than of decomposition of the fixed alkali salt. Even when the fixed alkali salt retains a little of the earthy part, that is of little consequence as we shall see below.

Note P
CALCINATION OF RED POTASH NATURE OF THE BLUE IN WHITE POTASH. POTASH DECOMPOSES ON DISSOLUTION AND BY CALCINATION

Not all red potashes are equally easy to convert into good white potash. Those of which the lye has not been clarified with care are the most troublesome. The gross colouring agent is more firmly adherent to the earth which remains in these potashes than to the fixed alkali salt. Next comes the red potash extracted from new ashes and then potash derived from oak ash. *Red* potash from the ashes of beech and of fruit trees is, other

---

a. The ordinary soda of Languedoc sells this year [1780] at from fifty sols to three pounds a hundred weight.

b. See p. 101 of the *Art of Glass*, French translation, quarto.

things being equal, the most easily converted into good white potash. If it is claimed that the blue tint taken on as a result of calcination of the potash is due to iron being converted into Prussian blue [ferric ferrocyanide], I cannot be of this opinion. 1. Because the Prussian blue can never resist calcination and that the fiercer the heat the bluer the potash becomes, acquiring greater intensity. 2. Because the acids that brighten Prussian blue make that of the potash disappear. 3. Because the blue of the potash is conserved on vitrification but that Prussian blue thereby becomes yellow in colour, which is always that of vitrified iron.

When the potash is very dry and has lost as much as possible of its yellow colouring matter under the action of the flame, direct contact with a flame is sufficient to develop the blue colour. If this is done before the yellow colouring agent has been entirely dissipated, the colour of the potash is more or less green because the blue is combined with the yellow. As the potash becomes purer and more perfect it becomes a better blue.

However the potash continually loses its blue colour as the duration of calcination is increased, it is even possible to bring it almost to the point of whiteness. What can be the reason for this phenomenon? After searching for it a long time, I believe that I have discovered it.

It is that the fixed alkali decomposes in dry conditions just as in wet; that its constituent part which it loses in either manner is intimately bound with the blue colouring agent and thus carries it away with it. It is impossible to develop any trace of blue in the salt precipitated from solutions of fixed alkali or in that remaining after a long calcination.

Fixed alkali, by this decomposition, has not only lost its blue colour, it has also lost its fluxing or vitrifying property. The constituent which carries away the blue colouring matter thus is the fluxing and vitrifying part of fixed alkali. It does not seem to me that one can doubt it. In whatever proportion one mixes this precipitated and well-washed matter from a solution of fixed alkali with flint-like matter, one never obtains any vitrification.

For a batch to have a convenient degree of fusibility it is necessary to use a greater proportion of white potash. Master glass-makers often make this increase without realizing it. They put in the batch as many pounds of white potash as they would use of red potash to obtain a good melt, although white potash is always drier and contains less water than the red potash.

NEW THEORY OF PRUSSIAN BLUE

It seems to me that what we have just said throws considerable light on the nature and formation of Prussian blue by calcination of one part fixed alkali with two parts of dried ox blood; the first takes up the constituent that it loses in an open fire or by dissolution and, consequently, also the blue colouring agent which is always very firmly bound to this constituent. When the solution of this fixed alkali is poured into a solution of green vitriol, two principal decompositions and two new constituents are evidently involved. Sulphuric acid having greater affinity for the ordinary fixed alkali than it has for iron, releases the latter in the form of bases, taking with it its *phlogiston* to combine with the former; the constituent of the fixed alkali which is in excess, having greater affinity with the iron compounds than with the new compound, the sulphated tartar releases the last to combine with the first and to form with its Prussian blue.

From this theory, which I believe to be the only correct one, it may be concluded:
1. That Prussian blue should be decomposed by ordinary fixed alkali, although it owes its colour to one of the components of the fixed alkali. This is, in effect, what happens. Whenever one digests Prussian blue with a solution of ordinary fixed alkali, ochre [ferric hydroxide] is precipitated.
2. That the beauty of Prussian blue depends only on the purity of the iron sulphate and on the alkali used being as pure and saturated as possible with that constituent part which the fixed alkali loses to an open fire or by dissolution. If iron sulphate is mixed with sulphates of copper and zinc the bases of these two substances, having no known affinity with the component that the alkali can yield, necessarily degrade the beauty of the Prussian blue. It becomes green partly from the yellow ochre mixed with it, if the alkali does not give up to the earth as much of its iron compo-

nent in excess, as it receives itself sulphuric acid, or if all the iron oxide precipitated from sulphuric acid cannot be coloured blue. It is certainly not easy to find copper sulphate which, at least, is not mixed with zinc sulphate, it is even possible that such has never existed. Good chemists agree without difficulty that nothing is less easy than the separation of these two salts.

3. That acids do not intensify the colour of Prussian blue because they combine with a part of the heterogeneous materials and alter the colour.

4. That there is no foundation for those chemists who make iron a universal colouring agent, who attribute all simple and compound colours to it in almost all substances, even flowers.

The iron oxide in Prussian blue is evidently only the secondary base of the blue colouring agent. It seems to me that iron plays the same role in all these cases but with different degrees of affinity. By calcination and by exposure to the fumes from urine and decomposing matter from a midden, iron is reduced to an oxide which becomes red; in the air and on vitrification it becomes yellow. Thus the substance of iron can be and actually is, in certain circumstances, the secondary base of red and yellow colouring agents, just as it is of the blue in Prussian blue; but it has less affinity with the secondary blue colouring agent than with the red and less with the red than with the yellow, since the blue is dissipated by calcination and the red by vitrification. The yellow of glass and iron oxide itself disappears on cementation. It is the same for the colours of glasses containing other metals, except for the purple colour of gold. Thus the purple colouring agent of gold is the most stable of all.

5. That the fixed alkali can become supersaturated with the component which it loses to an open fire and on dissolution; that it can, I say, become supersaturated by wet methods, being digested with a material that contains it, and by dry methods, when, well mixed with certain materials, it is exposed sufficiently long to the appropriate action of heat. The formation of Prussian blue proves that metallic oxides have the same power but to a lesser degree. May it not be that, by this property, fixed alkali, metallic oxides, and alkaline earths become fluxes for other compounds

that are infusible by themselves? Pure clay and the pure constituent of flint either separately or together do not melt at all in the most violent heat. If one mixes them with an appropriate proportion of fixed alkali, metallic oxides, or alkaline earths, vitrification is easily achieved.

6. That the names *soapy alkali*, *phlogistic alkali* and *animal salt*, introduced into chemistry in discussing Prussian blue, prove that there has not been a true understanding of this discovery. It would seem to me simpler and more conforming to justice to give to alkali supersaturated with the component which it loses by exposure to open fire or by dissolution, as has been done by Morveau, the name Prussian alkali. This would be a new way of marking our recognition of the authors of this precious preparation.

Some writers of very great repute think that the fixed alkali in the batch only serves to extract the glass which is already perfectly formed within the sand. This assertion does not appear tenable to me; ten thousand experiments, at least, have convinced me that the fixed alkali salts are no less constituents of the glass than the flint or quartz, etc. Every time that I have charged into the furnace two hundred pounds of sand and one hundred pounds of potash, or soda salt, I have obtained about two hundred and fifty pounds of glass; more or less according to the purity and dryness of the alkali.

* * * * * * * * * * *

PURIFICATION OF ALKALI

It is important to remove the principal colouring impurity as far as possible but repeated dissolution, filtration, and evaporation cannot achieve this. Infinitely better results can be achieved in less time and without loss by calcination, by exposing all parts of the fixed alkali to a bright clear flame until it has, when cooled, a bluish colour. The calcining furnace for the fixed alkali, of which *Kunckel* gave a plan in his remarks[a] upon *Merrett's* notes may be used, but it is very difficult not to say impossible, to avoid

---

a. See the *Art of Glass*, French quarto transl. p.319.

On Glass Making

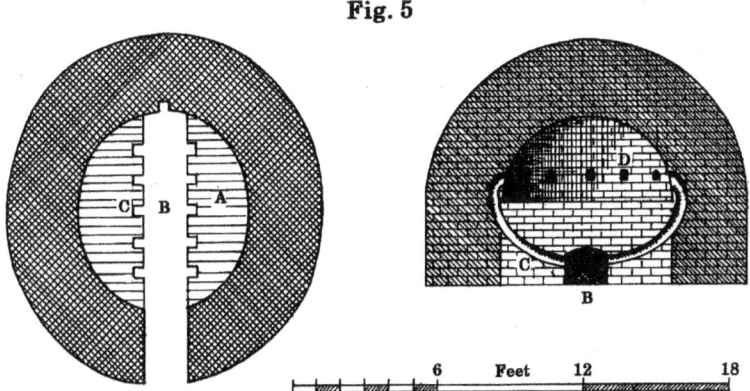

Figure 5. Plan and section of a furnace for calcining fixed alkali and making frit
A: Plan and shape of the hearth of the furnace; B: Fire box; CC: Flues by which the heat from the fire box is conducted into the furnace. Section: B: Stoke hole; D,E: [no explanation given].

ashes and wood charcoal being mixed with the alkali by the teaser, and that he will not lose a certain amount. That of which I append a plan (Fig.5) does not suffer this inconvenience and I can assure you produces the same effect with half the quantity of wood. Once heated up, one can in twenty four hours calcine about five thousand [pounds] of fixed alkali; four or five hundred pounds in two hours with one cord of charcoal or firewood. The only attention necessary is to stir the alkali well with an iron *rake* to prevent aqueous fusion which would retard the operation.

The fixed alkali salt of soda has an advantage over potash; it attracts moisture from the air less rapidly: master-glass makers can store it for a long time in appropriate wooden vessels without fearing any change and without needing to calcine it anew. To keep potash dry in these vessels it is essential to tar them on the outside.

Note Q
CONCERNING MANGANESE.
CONJECTURES ON ITS NATURE AND
COLOURING CONSTITUENT

It is customary and conforms with reason to give a mineral the name of the material that dominates in it, or of the most precious metal that it contains. Contrary to the announcement of Pott that iron is only found in manganese by accident, mineralogists continue to ignore this rule by putting manganese among sources of iron. It would be much more tolerable to regard it as a source of cobalt because there is never manganese which does not contain it and sometimes in large proportions, such as that of *Sivrac* in Rouergue, but one can discover no trace of iron in others.

Sage[a] seems to have carried the examination of manganese much further than Pott. He was the first to obtain, by distillation with sulphuric acid, crystals of zinc sulphate and obtained this semi-metal from the decomposition of these crystals by fixed alkali and distillation of the precipitate with powdered carbon.

It is undoubtedly a pity that Sage did not carry his discovery to the most complete demonstration; that he did not convert copper into brass using either the precipitate or the zinc that he obtained from the manganese.

It would be very useful to know whether zinc predominates in manganese, whether it is a source of zinc, but this would not instruct us any further about the nature of the red colour which manganese always gives to fine white glass. To soundly attribute the red colour to the zinc that it contains one would necessarily, it seems to me, have to show that zinc also colours glass red when added to it. In whatever proportions and whatever manner I have made this addition of zinc, either to the batch or to the melt, I have been unable to obtain a red colour.

Gahn and, following him, the expert chemists of Dijon have obtained from manganese a very singular spangled white metal which is neither

---

a. See vol. II p.134 of his *Elements of Mineralogy*.

iron, nor zinc, nor any other known metal or semi-metal. It is for this reason that Bergman did not hesitate to consider manganese as a source of a new semi-metal.

If one separates from manganese all the cobalt that it contains, it no longer gives a red tint to white glass. The red colour that manganese gives to glass is not bright, it is more or less violet. From these two observations should one not suspect that the colouring agent of manganese belongs to a part of the cobalt, modified in a particular fashion by a very strong acid which appears to have some of the characteristics of hydrochloric acid but that it is much more strongly fixed? I refrain from saying so; we have not advanced sufficiently far; this matter demands and merits new researches.

Manganese makes molten glass bubble and boil when it is mixed in before being vitrified. It has this in common with the fatty (rich) matter in lean red potash and all materials saturated with too much phlogiston.

The glass requires less manganese the better refined and less coloured that it is. That which is full of bubbles, of poor quality, milky, requires a large amount before becoming tinted red. Milky glass even becomes bluish more readily than it becomes red. This is because the blue colouring agent of cobalt has, so to say, enormously less affinity with the gall than the red colorant.

The manganese red is a fine colour but it never combines with deep yellow; one can mix manganese well with a very yellow glass, as much refined as possible, and this glass will never become red. The same effect can be observed in the art of dyeing.

The best manganese, the purest and the most strongly colouring that I have used, is that obtained from the Black Mountains not far from Sainte-Marie aux Mines.

* * * * * * * * * * * *

ON LIME

Few people suspect that it is necessary to add an alkaline earth to sand and fixed alkali to have good and beautiful glass. It is this, however, which procures perfect mixing, gives *body* to the glass, makes it stable and facilitates purification. Sand and fixed alkali give, unless one has too much of the latter, the melt a composition which is too pasty [viscous] for the materials to mix together intimately and for the gall to be dissipated and carry away with it the colouring impurities. One of the best alkaline earths that one can use is lime quenched in water and very white. According to Kunckel[a] chalk has long been used in the German glass houses. Lime has the same effect in the glass as the alkaline earth from vegetable ashes which has been made perfectly white; either gives the glass greater fluidity when molten and greater stability against yellow colour if used in [sufficiently] large proportion. Lime and the earth from vegetable matter bring much less colour to a glass the longer they are calcined in a very clear flame.

ON CULLET

I do not know any author who has spoken about cullet or broken glass, and master glass makers have condemned it to be used only in batches for ordinary glass. Broken pieces of the best crystal or of white flint are used only in batches for greenish glass or ordinary colourless glass. This prejudice is very damaging to the art of glass making: it inevitably increases by a quarter the prices of crystal and fine white flint. Cullet, far from impairing batches of the same type of glass, is beneficial to them, facilitating purification and producing crystal or white flint which is clearer, more stable and more brilliant. It is surprising that this truth has not been grasped: it seems to me the obvious consequence of the principle generally understood that crystal becomes better and finer the longer it has been heated or if it has been quenched many times in water. The only precautions required in the use of cullet are: 1. to remove all

---

a. See p.101 in the French quarto translation of *The Art of Glass*.

foreign matter, scale from the blow pipe, earth, tear drops, stones, etc.; 2. to crush finely; 3. to use only cullet of the same type as the glass being melted; 4. not to use more than one third cullet in the batch; 5. to mix the cullet well with the other materials. [Note R fits best here and S later]

Note R
THE CUSTOM OF CALCINING CULLET IS PREJUDICIAL

The habit of fritting the cullet with the batch, even of heating it to redness to quench in water, is harmful with all types of glass. There is none which does not become distinctly more difficult to melt the longer it is exposed to a calcining fire. The crude glass of some bottle factories will become completely unmeltable by the same. In *fritting* furnaces it loses a part of its vitrifying agent, just as it loses all of it on sintering with lime. It is consequently necessary to use a larger quantity of flux in batches in which the cullet is calcined than in those where the cullet has not been calcined. However new and perhaps odd this observation may appear to be, it is no less well founded or important.

All master glass makers could easily make themselves sure about this and perhaps, to convince them that it has been verified, it may suffice to remind them that after normal refining the glass becomes harder and more intractable the longer it is held molten in the pots; bottle glass develops a skin, losing its transparency and its fluidity; and that cullet alone cannot produce glass as fine, as transparent, as mellow, as easy to work and cut, as that made from batch.

It would be imprudent to mix cullet with batches which also contain red potash or badly fritted soda. The glass will always be markedly yellow when a large quantity of cullet has been used in the batch. One can, without any fear of inconvenience, throw the cullet into the pots after the colouring impurity has been dissipated or, much better, onto the white gall on the glass, if any is present during melting.

* * * * * * * * * *

DECOLORIZING WITH MANGANESE

In batches of the three raw materials mentioned above [sand, fixed alkali and alkaline earth], the blue colour always given by the fixed alkali combines with the yellow colour given by the lime or alkaline earth of vegetable origin producing a more or less marked green according to the proportions of the materials and their content of colouring agent. People are generally persuaded that *manganese* purifies glass, that it acts as a *soap* or destroyer of the green colour, but I doubt that this can be so following careful examination and reflection. The green colour reappears as soon as the red colour of the manganese is dissipated and I have never seen that the glass has gained anything. It seems to me more natural to think that the combination of the three primary colours blue, red and yellow, gives white in the glass. If the lime predominates or the batch materials have not been well-purified, the glass is yellow or yellow-green: if the materials have been carefully purified and if the fixed alkali predominates over the lime, the glass will be blue.

It is just the same with the *manganese:* the slightest hint of its red tint, all other things being equal, gives the most pleasing whiteness to the glass. I have confirmed by repeated experiment that, however much it may be recommended, slaking the manganese in vinegar adds nothing to its effectiveness[a]. Solid *manganese* from Piedmont is the best that is known. If *Caesalpin, Henckel, Linnaeus* and *Wallerius*, &c. had had an opportunity to examine this they would not have included it amongst sources of iron[b]. One can find in the *Miscellanea Berolinensia*, 1740, an excellent memoir on this material by M. Pott.

The materials that are commonly added to batches made of the constituents we have just discussed are not at all necessary. There is only *zaffre* which can be useful to prevent a yellow colour in glass made with poorly purified raw materials or with too much lime.

---

a. See p.53 in The Art of Glass, French quarto translation
b. See p.48 in vol. I of the *Mineralogy* of Wallerius.

BATCH COMPOSITIONS AND GLASS MELTING

Nothing is less uniform among the various authors or the glass houses than batch compositions. Each master glass-maker[a] has his proportions and his recipe. In the glass houses of *Bayel* in Champagne and *Novion* in *Tiérache* they use equal parts of sand and *potash;* in those of *Etembac* in the *Vosges,* equal parts of sand, lime and *potash.* How can so great a difference arise? Probably because M. *Drulanvaux*, to whom we owe the establishment in France of white glass made in the Bohemian style, has a melting furnace which gives a good heat but Mr X and others have furnaces which give only a feeble heat.

Having nothing fixed about the design and construction of the furnace, the nature and preparation of the new materials, on the true purification of the glass, it was impossible to have anything fixed and satisfactory on the composition of the batch. Even experience may only throw us into new obscurities or confirm us in our mistakes. Mr X maintained that his batch was good because however little he decreased the proportion of *potash* or *salt,* it melted badly or could not be worked. The batch composition of Mr X was good relative to his furnace but was very bad in itself because the glass which resulted was not very clear or brilliant, not stable, susceptible to humidity, sweated, and eventually decomposed; the hollow feet of drinking glasses became filled, in the warehouse and without any obvious communication with the outside air, with a salt liquor or solution containing part fixed alkali, part neutral salt. One sees nothing similar in the glass of M. *Drulanvaux*, without doubt because his furnace gives a sufficient heat to produce, with a small proportion of flux, a mutual penetration, an intimate mixing of the essential materials, sufficient fluidity for the gall to rise to the surface and be dissipated, carrying away with it the major part of the colouring impurity. The defects of compositions

---

a. See the Art of Glass of Neri, Merrett and Kunckel and also that of Blancourt; p.177 of the Lithol. of Pott and p. 136, vol.V of the Elements of Chemistry of Boerhaave.

that are too *soft*, too rich in fixed alkali, will not be anything like as bad in glass houses that have good furnaces. The long continued action of a very violent heat will even make the majority [of defects] disappear. Crystal and flint will not lack hardness and sufficient stability. Yet good furnaces cannot correct the defects of too lean a batch. However long the most violent heat is continued, the mixing of the materials is always imperfect, the glass unequally solid, full of bubbles, susceptible to attack and remains difficult to work. It never has sufficient fluidity for the essential materials to become intimately combined and for the gall to be dissipated.

[1.] Equal parts of good soda, white sand or sandstone, cullet of the same type and five ounces of manganese per hundred [pounds]; [2.] equal parts of good sand, well calcined lime, white potash, cullet of the same type and four ounces of manganese per hundred [pounds]; [3.] three parts of very white and very pure sand, two parts of fixed alkali salt, either soda or potash, very pure, one part cullet of the same type, half a part of lime calcined with the greatest care and four ounces of manganese per hundred [lb], usually represent, the best compositions for ordinary white glass, white flint, and crystal, respectively, with a good furnace. I say *usually* because the soda and fixed alkali salts can contain varying proportions of neutral salts which never make an entirely homogeneous melt together with sand and lime; at least those that include sulphates and chlorides do not.

It is much to be desired that we should have a sure rule to direct us on such an important and delicate point. I can report a means of finding the most advantageous proportions of fluxes which has always succeeded perfectly for me. I make two or three pounds of batch in the proportions given above for sand, alkaline earth or lime, cullet and manganese but I decrease the proportion of flux. When I have found the point at which this batch, put in a small pot in the working hole, simply melts during the time usually needed to give good refining, I increase by a tenth the amount of flux and the glass produced has the qualities that I desire. It is stable, very clear, very brilliant and retains a polish very well.

Note S
CONCERNING BATCHES. EFFECTS OF LIME. CONVERSION OF GLASS INTO PORCELAIN; PHENOMENA OF THIS CONVERSION. PROPERTIES OF THIS PORCELAIN

I had been badly advised concerning the batches used by the glass house at Etembac or Saint-Quirin. The confidence that I had in the person who gave me the proportions gave me no reason to realize this when I wrote my memoir. It is my task to point out the error and correct it. One third of lime in a batch containing even the richest white potash will always give very bad effects. This batch behaves very well in melting, gives a transparent white glass, more stable and less sensitive to chilling than a glass in which less lime has been used. But this glass, when worked in a flame, becomes opal and milky. The same thing happens with glass made from a batch in which the same proportion of powdered white calcined bones, alkaline earth from ashes, or fusible spar has been used. See my memoir on *the false emerald of the Auvergne*.

This phenomenon is too unusual and too important for us to leave it without giving a reason but it is appropriate to make a few preliminary observations.

Lime, however white it may be and however long it has been calcined, contains an abundance of yellow colouring agent, which only develops its colour during vitrification of the earth with which it is intimately bound. What proves this incontestably is that whenever one mixes lime or alkaline earths with a rich potash glass, the glass has a yellow tint, at the first moment of melting, but then the gall which makes the glass milky, having a greater affinity with the colouring impurity than the glass itself, combines with it and dissipates it. If it should be necessary, metallurgy provides us with a new proof that lime takes up a great deal of the colouring agent, of phlogiston; that it alone reduces metallic bases in a reasonable heat. On this subject one may consult the *Mineralogy* of

Cronsted, p.16 in the French translation[a].

One can make milky chalk-glass with much less lime but this is not part of the batch; it is only mixed into the molten glass when it has already been refined. This glass is worked and formed into the desired shapes before the lime can become vitrified; it is simply very finely divided and spread throughout the glass, like the gall in a too-rich glass or like the tin oxide in white enamel. Also this milky chalk glass is more fusible, more fragile, more sensitive to sudden heating or cooling, and much more difficult to convert into true porcelain, than that made from batch containing one third of lime. But the most remarkable difference between these two types of glass is that, in the latter, all the lime had entered perfectly into the vitrified melt; only after being exposed to the action of a flame did it become free again or *devitrified*, if one may use this expression and no longer remained an integral part of the glass but a body foreign to it. It is sufficient to examine the glass under a hand lens to be convinced of this.

For this vitrified material to turn into lime again it must necessarily lose its vitrifying agent and it loses it more easily under the action of a flame than any other type of white glass although the latter retain for a very long time their transparency, the most characteristic property of glass: it is no less indispensable that this vitrifying agent can, in some circumstances, become volatile and that it is always less fixed in the vitrified lime than in other types of white glass.

We have seen above that earth of the type of flint and all types of glass lose some of their fusibility and thus of their vitrifying agent at the heat used for calcining. Our glass with one third of lime loses almost as much of it as the coarsest glass.

All types of glass, in losing some of their fusibility or vitrifying agent,

---

a. Lime obtained from the demolition of iron foundry furnaces constructed from limestone and which has sometimes been calcined for a year, often contains as much colouring impurity as that which has only been calcined three times twenty-four hours. This lime is slaked with great difficulty but can be very good according to the nature of the stone.

their transparency, lose some of it on being converted into porcelain by sintering. That which is the most beautiful, the most transparent, loses the most on this conversion. Réaumur and, after him Pott, on the false assumption that the most subtle part of the lime enters into the glass during cementation and destroys its transparency, thought that the glass, far from losing weight during sintering, gains it. But this result is no more soundly based than the principle on which it is based. Experience has proved that to me many times.

It seems to me to be granted without difficulty that phlogiston is the agent of all volatility and that the vitrifying fluxing agent is found in the flame in a state of great volatility.

It does not appear to me difficult to provide a reason for the phenomenon. The vitrifying agent can only be firmly fixed in vitrified lime because of the phlogiston that it conceals and prefers to conserve but it is not in a state of sufficiently great volatility to escape easily through the whole mass of glass in the pot. It is not the same when this glass presents very extended surfaces to the flame. Then the vitrifying agent of the latter communicates its volatility, through an infinite number of points, to the vitrifying agent of the lime which can then remove it much more readily than it would itself have been disposed to volatilize.

Lime should make up only one twenty sixth part of the batch for white glass in which the potash is neither too *fat* nor too *lean*, and only one twenty-first part of those in which the richest white potash is used.

FUSIBLE SPAR ADVANTAGEOUSLY SUBSTITUTED FOR LIME IN BATCHES

I advise master glass-makers who can easily obtain fusible spar, green, violet, or any other colour, to substitute it for lime in their batches. It will produce very good effects and, being fusible by itself, it can take the place of half of its weight of fixed alkali salt, or one can increase the sand in the batch by the same weight. This economy is too precious to be neglected. See my memoir on the false emerald of the Auvergne. The most usual batch for miscellaneous white glass is two parts of good sand and one part of red or white potash. Quartz, as we have seen above, needs a little less fixed alkali. One adds lime in the proportion as the potash is more or less rich. It is indispensable to mix manganese with the glass when the batch is made with red potash. It is much less dangerous to add a little too much alkali to a batch than to use too little. The fire can correct one but not the other.

LIQUOR OF FLINTS: FALSE EXPLANATIONS THAT HAVE BEEN GIVEN

A large number of chemists have written about the liquor [soluble silicate] made with one part of flint and three or four parts of fixed alkali salt: many have remarked that during the operation the quartz earth becomes alkaline;[a] yet no one has said, though this is nevertheless true, that this composition with an excess of fixed alkali can produce, after a long and violent heating, a glass as beautiful and stable and as little attacked by acids as the most perfect ordinary glass.

This phenomenon does not appear to me difficult to explain. Under the action of a long continued and violent heat, the excess fixed alkali, which is the most exposed, decomposes losing its vitrifying agent, which is one of its constituent parts, at least as far as it can be vitrified and make a homogeneous material with the substance of the flint. A moderate calcining fire is sufficient to decompose fixed alkali salt. Also one should always use about one tenth more white potash than red. Why should that be! Why should not the fire extract the fluxing agent from the fixed alkali when it will eventually extract it from all substances, even glass itself?

I do not see anything astonishing in the alkaline character that flint-substance takes on in flint liquor, it evidently owes it to the fixed alkali with which it has been mixed to excess. It is quite natural that the lye of flint liquor to which sulphuric acid has been added should, on evapora-

---

a. See Pott's *Lithol.*, French translation, part I, p.174.

tion, produce a sulphate residue. This salt is the necessary result of the acid added and the fixed alkali found in the liquor because it has not been subject to the action of heat long enough to be vitrified.

It does not surprise me that the substance precipitated from flint liquor has lost its fusibility. It is very probable that at the instant of precipitation the fixed alkali salt, which has a much greater affinity with the vitrifying agent, has taken away the little that it possessed.

Is it not also very straightforward that the substance precipitated from flint liquor should have very marked relations to the base of alum. It is certain that this base can be found in flints, that flints can be converted into clay and that all clays contain the earth of alum.

Since this was written Achard and Magellan have shown without doubt that alum and flints both have the same basic earth.

I thought that it was in the interests of both the art and truth to raise these misunderstandings. The chemists who have made them have the progress of science too much at heart, and have contributed to it too efficaciously, for me to fear that they will take offence.

MIXTURES PRODUCING ONE HOMOGENEOUS MATERIAL

The ability that vitrification possesses of making one entirely homogeneous material from mixtures of different substances is not one of the lesser marvels. This singular ability merits our attention for a few moments.

I have often made batches with the four types of earth, flint, lime, gypsum, and clay and all the metallic oxides. Such mixtures made with care and in appropriate proportions have always produced in a violent and long continued glass furnace heat, a well refined, bubble free, stable glass but it is hardly transparent because of the colouring agent that it contains; being generally black in the mass but yellow in very thin sections.

If one repeatedly mixes gall with this molten glass it will remove the excess colouring agent; it is even possible to achieve a water green and good transparency. This experiment shows that the colouring agent diminishes the transparency of the glass. See my memoir on the nature of *the electrical fluid*.

Could one not conclude, concerning the black and yellow colours shown by this glass before it had been mixed with gall, according to whether one sees it in the mass or in thin sections, that the black is not a particular colour but only produced in this case by condensation of the yellow colouring agent? It seems to me very likely to be true that the three primary colours either separately or together can all give black if they are condensed and supersaturated in the glass.

Our glass, composed of the four earths and metallic oxides, coloured or decolorized, sintered for a long time at normal calcining heat, loses its transparency, its fusibility, and changes into an extremely refractory body, not attacked by acids, a white porcelain, if one has not used precipitate of Cassius in its composition: if one adds a rather strong dose of that, the porcelain produced is a pale purple.

We have many very singular things to consider here.

1. The four types of earth and the metallic oxides, except that of gold, lose all their characteristic properties on vitrification or sintering.

2. All the metal oxides, even that of gold, become completely irreducible.

3. All these minerals and metallic earths are reduced to one and the same earth, one that cannot be attacked by any acids and is the most infusible known. May it not be the true primitive earth, the elemental earth? It is certain that it appears very simple, that art cannot demonstrate its composition. The earth with which it seems to have most in common is that of flint, but it differs at least in its much smaller degree of fusibility.

4. The indestructibility of the purple of gold by sintering, although the colours that all other metal oxides give to glass are destroyed completely in this way. Were not the *vases murins** so esteemed by the ancients a purple coloured glass porcelain?

This suggestion does not appear to me to be entirely without founda-

---

* [*murin* = *murrhine*: a delicate type of ancient glassware from the East or, possibly, ware made from fluorspar (Oxford English Dictionary)]

tion: According to the rather obscure descriptions that we have these vases were red, had neither the transparency nor the fragility of glass and withstood very well sudden heating or cooling. These properties agree perfectly with our glass porcelain coloured with precipitated gold and tin. But I must leave these researches to persons better instructed than myself in the arts of the ancients and their main achievements.

5. That metallic oxides and alkaline earths, like fixed alkali, lose more completely by cementation than by vitrification their property of becoming supersaturated with the vitrifying agent, probably for the reason that, by this operation, they are reduced to their most primitive state.

* * * * * * * * * *

FRITTING

Fritting is nothing other than calcining the batch and this calcination serves only to mix the materials and remove the colouring impurity. It is indispensable to frit batches when they have been made with natural soda but there is no need with other batches if the materials have been well calcined separately. The mixing can occur just as well outside the furnace as inside it. The custom of adding *manganese* to the glass after melting does not seem to me to deserve to be followed. It is rare that it does not produce red streaks and that the irons and tools needed to do the mixing do not change the colour of the glass. If one finds that the manganese simply mixed with the other materials, is not sufficiently fixed, one has only to incorporate it by fusion with sand and fixed alkali salt in the manner of *smalt* or enamel blue. Then one puts the finely crushed red glass into the batch instead of the manganese. Frit and batch do not gain anything from being kept unless they are calcined a second time.

Note T
ANIMAL [PHOSPHATE] GLASS;
THE FOURTH TYPE OF GLASS

Henckel has reduced all artificial glasses to three types: vegetable glass, mineral glass, and mixed glass. If he had known the discovery of Rouelle, he would have added a fourth type, animal glass; thus each of nature's kingdoms has its glass.

M. Rouelle, demonstrator of Chemistry at the King's Garden, had been led to this important discovery, as he himself said in his *Chemical observations* inserted in the *Journal of Medicine* in October 1777, by those that had been made by Scheele concerning phosphoric acid in animal bones and stag's horn. This skilful chemist added to the procedure of Scheele and made it particularly interesting.

To obtain phosphoric acid from bones or stag's horn one calcines them until white, reduces them to powder then dissolves them in nitric acid; to this solution one adds sulphuric acid, then takes the liquor, removing as far as possible all the selenite ($CaSO_4$) which is found in the bones or which the sulphuric acid forms with calcium carbonate. This separation is achieved by means of a filter on which the selenite remains and from which all the phosphoric acid is removed by washing many times with distilled water. Then the liquid is evaporated in a double boiler and reduced to one quarter. The remainder is transferred to a luted glass retort with a receiver. Distillation then proceeds with an open, carefully controlled fire, until the retort is almost entirely red hot; when the interval between the fall of successive drops is 30 to 40 seconds one stops distillation. One finds in the bottom of the retort a hard whitish vitreous mass, more or less opaque, with an acid taste, which takes up moisture when exposed to air.

To separate the phosphoric acid and convert it to transparent glass one makes this mass dissolve in distilled water, reduces this to half in a double boiler, adds ten or twelve parts of spirit of wine to remove all the sulphuric acid and water that it contains. Then the phosphoric acid is precipitated to the bottom of the vessel in the form of a glue which sticks to the fingers. If one takes the material in this state in a good crucible and repeatedly heats it in many stages in a moderate heat, a stable, insoluble glass as transparent as crystal is obtained.

This glass, reduced to powder and mixed with powdered carbon, gives genuine phosphorus on distillation.

Rouelle noted that stag's horn gave more phosphoric acid than bones and that ivory, crayfish eyes, and mother of pearl gave none or extremely little.

This important discovery provides material for a large number of reflexions on the nature, formation and diseases of bones but we must here confine ourselves to those concerned with glass making.

1. However astonishing the phenomenon may appear to be, this discovery gives us obvious proof that phosphoric acid from bones is sufficiently fixed to withstand a glass making fire.

2. It is no less certain that this acid is a constituent of animal glass, that it is there even very abundant, since this glass, ground and mixed with powdered carbon gives phosphorus without itself decomposing.

One can convince oneself that this is not a true decomposition, that the glass has only lost its excess phosphoric acid, by a new fusion after separating the fixed alkali and the alkaline earth from the powdered carbon with a mineral acid. This second glass is even, if that is possible, more stable and more brilliant.

3. This last operation alone would prove that what Rouelle claimed for this glass is really true: it has all the essential properties: viscous melting, transparency, insolubility in all mineral acids, hardness, brittleness, and M. Proust, distinguished apothecary-major, who gained his skill at the General Hospital, has converted it, at my request, by cementation with lime, into porcelain.

4. It is very likely that our animal glass is not composed only of phosphoric acid, that the latter is only one of its constituents and that the flux is combined with some material which provides the base. This material does not seem to be soda, as Sage believed at first sight. Proust has made

extensive investigations on this subject and has assured me that he has never found it in fresh and uncalcined bones. Soda would necessarily be decomposed by the calcination, dissolution and washing to which the bones have been subjected; we have seen the reason above. It is thus very probable that the phosphoric acid has retained part of the calcium carbonate from the bones, or part of the base of soda which has decomposed.

5. Animal glass is a clear green. No one can ignore the fact that this colour results from a combination of blue and yellow: we thus here find again the yellow that alkaline earths always give and the blue that is given by the part of fixed alkali which is lost in an open fire or by dissolution.

The scientist Schlosser has obtained from human urine an acid very similar to that obtained from bones and which seems suitable to form a glass similar to that of M. Rouelle. See the excellent memoir of Schlosser in the Supplement, vol. 13 of the *Physical observations*.

* * * * * * * * * *

MELTING OF THE BATCH

Before putting the batch into the pots it is important that the furnace be very hot. If the heat penetrates the batch unequally it will melt less easily. The method of removing the *gall* during the first melting appears bad to me. This salt assists the melting and purification of the *second* charge. I have always found it useful not to make the second or the third charge until there are hardly any more bubbles in the proof drops. The glass then purifies itself much better and more quickly. If the gall is very abundant on the surface during the last melting, it is advantageous to remove it with an *iron ladle* because it corrodes the pots. *Gall* is the cruelest enemy that the master glass maker has to fear. It is the cause of bubbles, seed, white specks, congealing, blisters, threads, cloudiness, streaks, milkiness, cleavage, humidity, sweating, lack of stability, surface deposits, unequal attack, and sometimes fracture of the pots. It would be a very good thing if none would remain in the glass but I believe that it would be very undesirable if there was not a little in the fixed alkali salt. It helps the materials to melt, it facilitates their perfect mixing, contributes enormously to purifying of the glass, entraining with it the heterogeneities, especially the colouring impurity. To convince oneself one has only to make a melt with spilt glass which, through remaining in the bottom of melting furnaces, has become black and opaque. By this means one can give back its transparency and its natural colour and can even remove from it the electric property that it possesses to a very high degree. One can see from this that the custom in some glass houses of adding sodium chloride to batches for common impure glass is not to be scorned. Glauber's salt and sulphated tartar will produce the same effects; the last even deserves to be preferred because of its greater fluidity. See pp. 175 and 176 of the *Examination of Stones* by Pott.

Scientists and master glass-makers regarded *gall* as a *superfluous alkali salt*, before the excellent memoir of Pott already cited. They did not even suspect its large number of bad effects and could only be advised to employ two means of destroying it, *casting into water* or *very prolonged refining*. These methods are good but experience proves them to be insufficient. The water can only take up all the salt if the glass is reduced to a very fine powder by casting into the water. However long the action of heat may be continued, it cannot dissipate all the gall unless the glass has an appropriate fluidity. In such a case the glass retains the gall, just as too viscous slag retains the metal, the only difference is that one retards evaporation and the other precipitation.

Various glass house practices invented for other purposes contribute to the dissipation of gall. Some put *arsenic, antimony,* or *green tree bark* into the molten glass. Some stir it with pieces of green *ash, hazel, lime,* &c. with the idea of bleaching or destroying too strong tints. These different procedures are only effective if gall is present. They facilitate its release by the passage that they open through the melt: with arsenic by the paper in which it is wrapped, with antimony and bark by the water that they and the bark on the pieces of wood contain. The gall follows these materials and also carries colouring impurity away with it. The rod made to mix manganese into the glass has the same effect. After these operations the glass in the upper part of the pots is often full of bubbles and contains a half more gall than that in the middle or the lower part of the pot. One can confirm this by experiment.

These means never fulfil their object completely. One can only expect to destroy the gall completely if the batch has been made in appropriate proportions and that one can use a very violent heat sufficiently long continued, and that the batch contains the fixed alkali salt needed to *saturate* completely, if one is permitted to use this term, the sand and lime in the very violent heat. Perfect purification will then take place without having recourse to any other method. I believe that I have indicated these proportions above or the true method of finding them in all cases.

According to what we have just said, it can be seen what should be thought about what authors on the art of glass making have written concerning colours in glass and their being *consumed by fire*; that they can be evaporated; that minerals alone can supply them; &c.[a] The colours only disappear because the colouring agent has greater affinity with the gall than with the glass and becomes combined and dissipated together with it. when the glass is correctly purged of neutral salts, *sulphated tartar, sodium chloride* and *Glauber's salt,* the colours remain fixed in the most violent and long continued heat. The yellow colour given by soot, vegetable or animal carbon, is as well fixed as the blue of *Zaffre* and the red of *manganese*. Gall seems to me to be the surest means that one can employ to adjust colours to the tint and intensity desired.

FUELS FOR GLASS MAKING

All types of wood are suitable for melting and annealing glass provided that they are well dried and of a medium size, three to four inches in circumference. The wood of *beech, hornbeam, birch, cherry*, and *ash*, without bark, give the clearest flame: they are to be preferred, above all during working of the glass. The ashes of *poplar, aspen, willow* and *lime* are so light

---

a. See pp. 254, 255, 262, &c. in the *Art of Glass*, quarto translation and p. 275 in vol. V of Boerhaave's *Elements of Chemistry*.

that they fly about in the furnace and stick to the surface of objects being heated in it; this impairs their smoothness and brilliance. Oak, unless it is extremely dry, crackles and these small explosions throw particles into the pots. There is no fuel which has as sure and rapid a heating effect as good coal. The English crystal factories and the plate glass manufacturers of London and of *Tourlaville* in Normandy use it with great success. I have always found that forty pounds weight of good coal produces as much effect as twenty-five pounds of dry wood. There is certainly no need to fear that, in a well proportioned furnace, the smoke and ashes of coal will alter the colour of the glass. We have excellent coal in abundance [whilst] wood becomes day by day rarer and more precious. I believe that one cannot give too much encouragement towards the establishment of glass houses which burn only coal.

Whatever fuel is used, it is essential to feed the furnace very uniformly. The least negligence on the part of the *teaser* will considerably retard refining. When, for about an hour, no more bubbles have appeared in the test dips and the white flint or crystal is of the colour that one wishes, the fire can be stopped. It is important to stopper the working holes carefully and to keep them stoppered for three or four hours. One neglects this at one's peril; it contributes greatly to the perfection of the refining, not by giving the glass the ability to chase the air out of its interstices as it sinks down and becomes free from bubbles, as is commonly believed, but by allowing the foreign materials, principally the gall, which the glass may still contain, time to rise to the top of the pot. [Note U fits better later than here].

It would be almost useless to make extensive observations here on the methods of working the glass because this would take us much too far astray. I content myself with recommending great tidiness to the workmen.

ANNEALING

However important the annealing of the glass may be, I know of no author who has discussed it. In the glass houses they have extremely obscure ideas about it, there are even those who believe that it occurs by a type of occult virtue. It is easy to assure oneself that it is nothing other than cooling brought about by insensible degrees but it is very difficult to cool glass in this way, above all with pieces of unequal thickness or in large panes. Without doubt this is the reason why the sheet glass produced by the five large glass works of Normandy is badly annealed. However thin the sheets may be, they hardly ever take the scratch of a diamond properly and are nearly always more fragile than is the nature of glass. The method by which large pieces, lamp glasses, large pieces of optical glass, plates, etc. are usually annealed can only produce bad annealing.

When one has put into a very hot furnace all the pieces that one wishes to place there, it should be sealed up with care. It is certain that the glasses then receive, and almost at once, a greater degree of heat after sealing up. It is sufficient to look into the furnace to be convinced of this. Besides the breakage and the sagging that usually results from this increase in heat, one has to wait a very long time for the complete cooling of the furnace; at least according to generally received usage which is that one does not begin to open the furnace and admit air until the end of two or three days. The warmer the furnace the more rapidly does the outside air rush in. It is easy to see that opening up the furnace can easily subject the glasses to a greater change than they can withstand. Many pieces are always broken and all of them badly annealed. There is a very simple means of obtaining good annealing in a shorter time in the same furnace whilst avoiding these inconveniences ruinous to the master glass-maker and very prejudicial to the interests of the public. The only change is to make one or more holes five inches in diameter, according to the size of the furnace, in the middle of the furnace crown. For example, one hole will suffice for the furnace used to flatten and anneal blown flat glass or Bohemian style glass. It should be in the middle of the crown A (see fig.

6, p. 187). As soon as one has sealed up the openings F, G, and the flue C, one opens the hole in the crown and then closes up both ends of the hearth. In this way neither sagging nor breakage is to be feared; the glass will be annealed as perfectly as possible and in the space of four or five days instead of the eight, at least, needed for bad annealing. It can easily be seen that the degree of heat cannot increase and that it will decrease very slowly; that heat is lost from the hole in the crown without the outside air being drawn into the furnace to any appreciable extent.

Note U*
SOME IMPORTANT OBSERVATIONS RELATED TO ANNEALING

We believe that our memoir has given the most correct ideas about the annealing of glass. what we have to add is only a development.

It is impossible to anneal glass in the pot in which it has been melted, or when it is stuck to another body, without it cracking or breaking. For it to remain in one piece it is necessary that its contraction be exactly the same as, or less than, that of the pot or the other material to which it is stuck. Precisely the contrary is found; anyone who has seen the casting and cooling of plate glass will be convinced that of all substances glass contracts the most in volume during cooling.

I believe that I have observed that not all types of glass show the same contraction. The finest, the most homogeneous, has always seemed to me to have the advantage. May this not be because it contains the most vitrifying agent, which is subject to a greater expansion than the earth base or bodies foreign to the glass?

Whenever the glass is more fragile than is in its nature, it is *sour*; if it is not easily marked with a diamond and the crack will not follow the scratch, it is usually said to be badly annealed. [However] this bad quality is not

---

* [Although the text here refers to Note W, Note U is better here, before W, than earlier.]

always the result of poor annealing. We have already remarked that glass exposed too long to the heat and that made entirely from cullet, or with too much cullet, is difficult to work and to cut. That in which too much alkaline earth, or too little fixed alkali salt, has been used also becomes *sour* if heated briskly, and intractable especially when cold.

Master glass-makers who make ordinary sheet glass on one side and glass for goblets on the other, in German style furnaces, have had the sad experience. This sheet glass will never follow the diamond scratch in cutting although the same batch will give an acceptably sweet glass, easy to cut, when made in a furnace, intended only for this type of manufacture or in one making cast glass on one side. The reason for this difference is very simple; in furnaces for ordinary sheet glass and cast glass the heat is very much moderated during working but, in furnaces for German-style goblet manufacture, the heat is necessarily as violent during working as during melting and refining. Sheet glass containing a great deal of alkaline earth, when exposed to a very brisk and sustained flame during working *dries out* or *burns out* in the language of the workmen who observe it, loses a great deal of its vitrifying agent, and can never be easy to cut however much attention has been given to its annealing.

Master glass-makers who, in certain circumstances, find it necessary to practise this type of manufacture have a means of avoiding its most serious bad effect; that is to add to each hundred pounds of batch five pounds of good red or white potash. What we have just been saying furnishes us with proof that, in the useful arts, exactly similar effects do not always have one and the same cause. This is a fertile source of prejudices and errors, sometimes very serious, nearly always harmful to the fortunes of both seller and buyer.

* * * * * * * * * *

Note W
ON THE NATURE OF GLASS AND THE VITRIFYING AGENT

Glass is the most beautiful product of human industry; it is a body that is viscous when molten and when half-molten is susceptible to being shaped into every imaginable form; which when cold is transparent, transmitting, refracting and reflecting light marvellously. It is perfectly elastic and insoluble in the three mineral acids. These properties distinguish it from the four types of *earth*, salts, and metals.

The majority of scientists have believed that one can say that glass is the most fixed and most indestructible of all known substances. We take credit for having solidly proved that this opinion is devoid of all basis, that glass continually decomposes when kept hot, both during melting and during working, and that it is completely decomposed by sintering with another substance.

It is not only alterable by dry means, it is equally so by humid means. Marggraf had observed, before Scheele and Sage, that the the acid of vitreous spar [hydrofluoric acid] decomposed it and Rouelle noted that phosphoric acid from bones produced the same effect. Glass can only become, in reality, the most fixed and least destructible of all bodies by conversion into porcelain; but, in this state, it is no longer glass, it has lost its fusibility, its transparency, its elasticity, etc.

Most chemists, both ancient and modern, regard heat as the universal solvent, as the cause of all fluidity, as the sole agent responsible for melting and vitrification. There is no person who is not convinced that, without the aid of heat, one could not melt any metallic or earthy substance and this is a certain fact: but is this fact properly understood? Has it led to correct ideas? I do not think so. The scientists Morveau, Maret, and Durande, who preside over the public sessions of the Academy in Dijon, seem to me those who have come closest to the truth[a].

---

a. See p.174 of vol.1. of *Theoretical and Practical Elements of Chemistry*

To understand what we maintain on this important subject it is sufficient, I believe, to examine with care and to reply in full to the following question: is melting itself a necessary effect of the direct or indirect action of heat on earthy or metallic substances?

1. It seems to me that if fire were the direct cause of melting, all molten bodies, so long as they remained molten, would have the same properties. Yet liquid salts, metals and glasses have very different properties. The first two flow like water, the last can be drawn out, is viscous, etc. This striking difference, seems to me to indicate different direct causes for their melting.

2. If melting and vitrification, which alone interests us at the moment, were the result of the direct effect of heat, all substances exposed to it would be equally vitrified and one would have no need of the idea of fluxes, heat being the only one. However, one cannot ignore the fact that, however advantageous it would be to make glass from sand alone, without the help of other more expensive materials, this is impossible. If certain calcareous earths vitrify it is only on account of the natron that they contain and the vitrification of certain clays is only due to the iron oxides that they contain. For heat to cause vitrification it always has an indispensable need of fluxes; it is thus not itself the direct cause of vitrification; it is therefore not the true vitrifying agent.

3. If heat were the vitrifying agent, it would no less readily melt glass converted into porcelain than the batch initially used to produce it, &c.

One does not say that the fluxes only fix and retain the heat, preventing too rapid a loss, and thereby supplying in this manner a degree of heat which art alone cannot produce. It is agreed that the heat does not cause vitrification directly, but this is not the only property that the fluxes possess, it is not the only role that they play in the pots; they have another much more important function which is to dissolve the earthy substances and become intimately combined with them.

Not all the constituent parts of the fluxes fulfil their double role; we have already seen that when the fixed alkali salt is deprived of the constituent that it loses to an open fire or by dissolution, it cannot be melted or made into glass by itself and does not differ in any way from the alkaline earths. Thus it is only the constituent that the fixed alkali salts can lose which, because of its great affinity with heat, can retain and become supersaturated with heat, being thus itself diluted and put into very rapid motion, which then, because of its affinity with the earths, acts upon them with all its energy, both natural and acquired, dissolving them and making a viscous melt, a homogeneous body, a glass, by combining intimately with them.

It seems to me that one cannot doubt that there is both decomposition and formation of compounds: to prove it I believe that it is sufficient to recall what we observed earlier, that porcelain made from glass does not have the properties of any of the four earths, it is neither that of limestone, nor gypsum, nor clay, nor that of flints; it is very probably the primitive or elementary earth, at least it seems to me to combine all the properties of the latter.

The fixed alkalis and salts which can be made alkaline by simple calcination are not the only fluxes. One should place the alkaline earths and metal oxides in the same rank.

These three types of flux have characteristics which distinguish them from one another. [Fixed] alkali salt never forms glass by itself and appears to have the greatest reaction with the earth of flint. Metallic oxides can vitrify by themselves and appear to dissolve clays most easily. Alkaline earths never melt by themselves and attack clays the most vigorously.

It is not easy to find reasons for these differences and it will be much more useful to seek to understand the nature of the constituent part of the fluxes which produces glass and becomes a constituent of the glass produced.

The ancient chemists have removed for us some of the difficulties of this important research, saving us much trouble. These fathers of chemistry, too neglected in this century, and who often seem to have misled us, have honoured us with their real or claimed discoveries under various names. I may say that these pioneers have said, in very clear terms, that this vitrifying constituent of fluxes is an acid, *acidum pingue,* acid of flame or of heat, an acid very different from the three mineral acids, a spirit acid, invisible, insensible, &c. Tachenius had discovered this acid in the vapour of coal a long time before the Duke of Chaulnes had made his beautiful experiment. He also proved, with as much precision as modern workers have achieved in the last few years, that the increase in weight and hence the vitrifying property that metallic oxides acquire during calcination was solely due to an acid, the acid of fire[a].

The indefatigable Meyer has gathered together, related, refined, and extended the ideas of the ancients on this interesting subject. Before his researches on quick lime, he attributed *glass formation*, like all other chemists of our century, *to the power of heat*. He then acknowledged that he had been forced to give the honour to *acidum pingue*. He assured us that *acidum pingue is a true constituent of glass; that it is there found in its greatest concentration, that it enters into its composition and also the fills its pores, that it is the agent conferring its fusibility, &c*[b]. Sage has also had the same idea, see his memoirs on Chemistry.

The discovery of Rouelle, of which we have already given an account, seems to me to place beyond doubt the doctrine of the ancients and of Meyer on the nature of the vitrifying agent. The different names given to this acid of heat, son of the sun, spirit of acid, *acidum pingue*, of gas, of fixed air, acid of spar, phosphoric acid, do not alter its essential character; they only prove that it is not to be confused with any of the three mineral acids and also that chemists have observed it in many different circumstances, thought about it from many points of view, and in many different modifications.

* * * * * * * * * *

---

a. See ott. Tachen. clavem. &c. *Tract. de morb. princip. and hippocrat. chym.* cap. XXVII.

b. See pp. 207, 209 and 210 of the second vol., French translation.

WASTE OF LABOUR AND HEAT IN GLASS HOUSES

In glass-houses making flat glass, cast or blown, crown or cylinder, lamp glasses, crystal, &c., the workmen have too long intervals of rest. The majority are at least thirty hours on end and three times a week without work. This idleness is harmful to good working. It is rare that it does not produce slackness, that it does not lead to dissipation and even debauchery. The public is as interested as the master glass-makers in seeing this vice eradicated. We could provide ourselves with means equally useful and fair.

A great deal of heat is wasted in these same glass houses. The working holes are almost always free, neither the arches nor the annealing furnaces are ever full and the former are often empty. It would be very desirable to profit from this wasted heat and one could do so. We have men of exceptional talent for enamelling but these great artists have to make and prepare their own colours and enamels for themselves, at great expense, or buy them from foreign countries. What facilities could be found for this purpose in our glass factories! The men, the furnaces, and the heat necessary to calcine tin or lead, or to make metal *crocus* if this is preferred to oxide of tin or lead in compounding white enamel, would cost nothing. One could there find admirable means of preparing all colours and of properly purifying and firing enamels. We could certainly have enamels at very low cost and of exceptional quality, of the same degree of fusibility because the same heat and the same batch could always be used. Painters could also probably acquire beautiful colours and a greater number of tints.

Note X
IT IS THE ART OF GLASS MAKING THAT PROVIDES THE TRUE PRINCIPLES OF OTHER MANUFACTURES NEEDING FIRE

This proposition seems incontestable to me. I venture to suggest that one can find numerous proofs of this in my works. What is certainly true is that if I have had a few insights, if I have made some discoveries in pyrotechnics, pottery, mineralogy, quantitative analysis, and metallurgy, I owe them entirely to the art of glass making. Is not the reason why this precious art has not been deeply studied and developed that the other arts using heat are in the state of the greatest imperfection? Is it one of which we have the true principles, the practice of which can be illuminated by the torch of a reasoned theory? Is it one which should not again be given over to blind routine? However one or two examples will demonstrate the sad truth more strikingly.

We have many works on the art of making porcelain. Nearly all chemists have spoken of this important branch of pottery, but to what can their writings be reduced? To the description of a few practices, recipes always related to the local circumstances of a few manufacturers and rarely applicable to all. One finds, I dare to say, no definite principle, nothing perfectly satisfying about the nature and compounding of raw materials, the preparation and application of the glaze; on the relations that it is necessary to have between the paste and the glaze; on the nature, preparation, and application of colours, on the nature and preparation of the flux to be used to apply them to the glaze to make them produce the most agreeable effects and prevent all risk of dissolution; on the matching of the fusibility that must exist between the colours and the glaze, and between the colours themselves; on the degree of heat needed to fire the porcelain, on the means of producing it, on the construction of furnaces, on the composition of the saggars, &c.

The proof that there is no exaggeration in what we have just said is that each manufacturer has his own particular recipe on all these points of the art; that no one can at present make flat ware, even plates, and that small hollow pieces are sold at an exorbitant price.

Chance led to the composition of porcelain as I have indicated in my memoir. But it should be understood that I wish to speak of the new porcelain, called hard paste. It is certain that the composition of the paste is found ready made in the beautiful white earth of St. Thiriey in Limousin; it provides the pure clay and the pure sand that I have advised. The glaze of sand from the earth itself, fixed alkali, calcined gypsum, and litharge that is applied to this paste only differs from my lime glass in the lead oxide. I do not think that connoisseurs and friends of good health find this difference to the advantage of the new porcelain.

What has been the result of this discovery? Nothing extraordinary. [Only] what one should expect from ignorance of the principles; mustard pots, cups, small coffee pots, never undistorted plates, no piece of flat ware deserving any attention, and all at a very high price. Could I not flatter myself that no one has gone beyond the limits at which the subject of my memoir compelled me to stop?

If this is a loss to the public, I venture to suggest that the means of making it good could be found in my works. My country has delayed too long in making an ordinary porcelain, more stable, less sensitive to heating and cooling, as easy on the eye, and definitely less expensive than that from China.

We do not have very large resources, and for the same reason, in the art of metal founding, the most important of all after agriculture, and without which agriculture could not supply our needs. We shall see, lastly, that nothing is less developed or more imperfect; that the majority of what has been written is more likely to perpetuate errors and prejudices than to help manufacturers make better iron and to do so more economically; that Buffon is the only writer on this subject who has touched the heart of the matter, &c. The proof is to be found in our memoirs: means of classifying all known irons; illustrations of manufactures needing heat, the art of extracting from all ores the purest iron as granules or *en masse*;

the art of converting iron into steel, all reduced to the greatest simplicity.

The advice that I have given master glass-makers to use the waste heat from their furnaces to make steel by cementation has not been as useful as I had imagined. I had supposed, following the celebrated Réaumur, that we had in France iron sufficiently pure to be easily converted into good steel. This, unfortunately, was a false assumption. I can confirm, without being contradicted by experience, that no foundry in the kingdom can produce iron sufficiently pure to be converted into fine steel by cementation. Such has, until now, been made in France and also in England only with the iron from Roslagen in Sweden.

* * * * * * * * * *

The art of the mosaic is the only one that can transmit without alteration, to the most distant posterity, remembrances of great events and illustrious persons. Its products withstand equally harm from the air, the action of humidity and the resources of malice or jealousy. There is no traveller who is not enchanted by the masterpieces of this type that may be seen in Rome. This art is almost unknown in France except by name. Why has the art not spread outside Italy? We certainly have many artists who could distinguish themselves in this medium: it is not the talents that are lacking, it is the material. If they were provided with the different coloured glasses needed you would see prodigies. These coloured glasses could undoubtedly easily be made in our glass houses and at very low price.

We will only be able to flatter ourselves that our porcelains have achieved the degree of perfection of which they are capable when many people occupy themselves with this important branch of glass making. I do not believe that our glass houses could undertake the large pieces, the merit of which lies in the elegance of form, the correctness of the decoration, the richness of the composition, the boldness of the brushwork, the harmony of the colours. These prodigies of art and taste are reserved to the factory at Sèvres but I believe that the glass factories could make a medium porcelain, sufficiently good and at a sufficiently modest price, to stop the ruinous importations of ordinary porcelain from China.

Pure clay, well prepared and properly combined with powdered very white flint or sand can provide a sufficiently good material for porcelain and a well purified lime glass can provide a good glaze.[a] This porcelain does not differ as much as might be expected from that of China; at least one should recall that it uses neither salts in the paste nor metals in the glaze. Master glass makers would need only a painter and a few models. They would find that, at no or very little cost to them, they had everything else that was needed; workmen, buildings, furnaces, and all the degrees of heat needed.

The conversion of iron into steel by *cementation* could also furnish glass makers with another means of profiting from their waste heat and of occupying their workmen without fatiguing them.

It is known that the art consists entirely in making the iron harder and charging it with a greater quantity of *phlogiston*[b] diluted by means of salts. This industry would have a double advantage for the kingdom: glass could be sold at a lower price and we would not be obliged to import, for very considerable sums every year, steel from foreign countries.

These means would allow profit to be obtained from the over long rest of the workmen and the heat wasted by the glass houses and deserve to be treated at greater length but I cannot stay longer without exceeding the limits prescribed for this memoir. I must give notice that I have not intended to confuse the gentlemen glass makers with the workmen. It would not appear just that they should not be better paid especially since they work as long as the ordinary workmen. Their natural desire to distinguish themselves will assure both the art and the glass works owners an ample recompense.

These would not be the only advantages that we could obtain by bringing the art of glass making to perfection. [However,] we must wait for a more intimate and extended knowledge of salts, phlogiston, colours, earths, minerals, metallurgy, pyrotechnics, &c. It would not be difficult

---

a. See pp. 103, 605 &c. of the *Art of Glass*, French quarto translation.
b. See the Art of converting iron into steel by the celebrated Réaumur.

to prove that this claim is not without foundation.

New insights could without doubt contribute to the perfection of the glass making but they would never lead to it without the greatest degree of help that the government could provide. The art is too precious and the Minister too clear sighted for us not to be persuaded that he will accord it all the facilities that it needs, that he will honour new researches with his protection, that he will encourage the industry and that he will inspire the talented. He will surely never permit me to give any details in this respect. I must conclude this in accordance with the wishes of the good citizen which gave rise to this memoir[a]. I believe that I will be pardoned for having passed rapidly over many points that I have ventured to consider. There is too great a scope to be covered in a memoir. [However,] I can testify that I have not neglected anything that I believed it very important to say. If I did not fear to become boring, it would be easy for me to establish by the state of expenses and of production that, by means of what I have said about furnaces, pots, raw materials and the manner of making it, good durable glass could be made and sold for half the price charged by our most esteemed glass houses. I could perhaps have avoided some trouble by applying my principles only to some particular branch of glass making for example, plate glass, but that would have led me much too far. If I could believe that this was a loss to the public I would not be backward in making good the omission; nothing would flatter me more than to be so encouraged by the Academy and to merit its suffrage of what I have the honour to present to it.

---

a. A good citizen who wished never to be identified gave a sum of five hundred pounds to the Royal Academy of Sciences in 1759 to reward whoever, in the opinion of this body, made the best reply to the question which provides the title of this memoir.

MEMOIR ON THE CAUSE OF BUBBLES FOUND IN GLASS

*Read to the Royal Academy of Sciences
at the beginning of 1758
and printed in the
fourth volume of the* Savants Étrangers*

THE ART of glass making is one of the most curious and most deserving to occupy true physicists, its phenomena are very remarkable and the utility of its products very extensive; nevertheless there is much about it which has not been understood as deeply as one would wish. The most esteemed treatises on glass making only give very imperfect ideas: one does not see in Agricola, Neri, Merrett, Kunckel, Henckel, d'Ablancour [de Blancourt], &c., virtually any principle solidly established or any phenomenon clearly explained. In these authors all is reduced to a few particular things, to methods and precepts related to the materials of the countries where they lived, and the furnaces by which they were served, and consequently are of little use to those who operate in different circumstances. They have nothing satisfactory to say about the material, preparation and construction of furnaces, on the composition and shape

---

* Elsewhere this memoir is said to begin on p. 555.

of pots, on the proportion that one must have between the pots and the furnace, on the most advantageous degree of heat, on the nature of the materials to be converted into glass, on the causes of purification, on transparency, colours, variations in stability, bubbles, cloudiness, smears, weathering or tarnishing, on the nature and effects of good annealing, &c. Also, in the countries where these authors are best known, and where nature seems to be most favourable to glass making, good quality glass is made only by chance and at great cost.

What can be the reason for glass making having made so little progress? I believe that I have discovered the reason; it is that the majority of those who, by their station or their interests, cultivate it, lack the insight necessary to develop it; also that the small number of people capable of penetrating its mysteries have not had opportunities to work on a large scale, the only means of discovering the true principles. In ordinary laboratories the marks of its nature are not susceptible [to observation]: in small glass houses these characteristics are too slender to be easily observed; it is in the largest, in the manufacture of plate glass, that these characteristics become striking.

I have found myself in very fortunate circumstances: if my insights and my talents had been in proportion to my good will and my efforts, I would have left little still to be done in this matter.

The Academy is the most competent judge of researches and discoveries that have contributed to the good of the State and to the progress of the sciences; whatever judgment it may wish to make on those which I shall have the honour to place before its eyes, I beg that it will regard them as a proof of my zeal and my sincere wish to merit its approval. If this memoir is agreeable to the Academy, it will be followed by many others.

All those who have treated glass making have regarded gall or salt water as a superfluous alkali salt; glass makers have the same idea of it. Few chemists have talked about it; Pott is, I believe, the first to have examined it with care; his researches have been crowned with great success. He has demonstrated that gall is not an alkali salt but a mixture of different neutral salts, Glauber's salt [$Na_2SO_4.10H_2O$], sulphated tartar [K_2SO_4]

and sea salt [NaCl]. The experiments that I have made myself on this salt have shown me nothing contrary to the findings of this expert chemist: his memoir which is to be found among those of the Berlin Academy very much deserves to be read.

The same author has stated in his *Lithogeognosie* that gall will not form glass in any fashion with vitrifiable earth and that it cannot enter in any way into the composition of glass. I cannot doubt the truth of this assertion: in whatever manner I have treated gall with sand I have never obtained the least appearance of vitrified material; I have always found the sand in the crucible without it having appeared subject to any change. This experiment demonstrates conclusively that gall is not a fixed alkali salt.

It is unfortunate that Pott did not push his researches much further, that he did not follow the behaviour of gall actually in the glass maker's pot or, much better, during the melting and refining of glass or in the manufacture of plate glass. What a vast field for an observer of this order! What a pity that he did not undertake it. Glass making must regret that he lacked the time and opportunity.

Gall plays a very extensive role in glass making; it has a great number of good and bad effects which I do not believe to have been suspected until now. Whoever understands these in detail will have one of the principal keys to the art of glass making. Such a subject could not be treated, I think, in too much detail; it could furnish me with material for many memoirs but I shall here restrict myself to one important effect of gall.

Bubbles, or blisters as they are termed by the glass maker, which are often seen in all types of glass, have always been regarded as the products of air. There is, it may be said, air everywhere; that which is in the glass is pushed towards the centre by cooling of the outside surface; it forms cavities which are almost empty, complete cooling having allowed it to condense: others assure us that bubbles only occur in the glass because the moment when it was good to work has not been seized; at the moment when stoking the fire ceased, they say, the mass was in considerable movement and this agitation must necessarily create interstices which the air rapidly fills. If the glass is worked before it has expelled the air in settling to its proper density, it is no surprise that there are bubbles in the products. I need not stop to remark on the weak foundation of this last explanation.

This cause has always seemed suspect to me; I have never been able to understand how air could be in, or be introduced into, such a blazing hot material, nor how it could be capable of carrying this effect to the point of rarefaction and shrinkage [in the place] where it should be, if it had been involved. I had in mind to discover the truth, above all since I have been myself especially involved in glass making: it might be necessary to search for a long time but I had the good fortune to succeed.

Towards the end of 1755 I had reason to believe that bubbles were the effect of a material much more substantial than air. I made a batch which refined very badly; the glass was full of bubbles of different sizes, although I had taken the greatest precautions, and it had been exposed for a very long time to the most violent heat. This phenomenon seemed very unusual and I thought it too important to find the true reason not to observe it again with as much attention as I could command. I made the same batch and everything that had happened before occurred again.

I had the pot which contained it removed from the furnace; on the surface of the glass there was a sort of skin in which innumerable bubbles could be seen; this skin was removed and a whitish vapour arose, which decreased as a new skin formed. There were no fewer bubbles in this new skin than in the former; this operation was repeated many times and I observed the same things, the bubbles and the vapour, every time: from that moment it seemed clear to me that this vapour was the cause of the bubbles. I was very impatient to discover its nature; it seemed to me to resemble very much the last fumes seen during melting, those which succeed the blackish and the reddish.

Despite the difficulties that arose, I collected and condensed a sufficient quantity of these whitish fumes to confirm for myself that they were nothing other than gall reduced to vapour. I made for a third time a melt of the composition described above and again saw the same phenomena; the scum was removed [repeatedly] and cast into water, until there was no more glass left in the pot. I had arranged that the vapours would be collected by a sort of cardboard bell jar well soaked with water. After this operation the cardboard was macerated in water: the next day the water was squeezed out forcibly, the liquor filtered and, having evaporated it, I found a very small quantity of gall. I also evaporated the water into which the glass had been cast; it gave me an ounce and a few grains of gall. If the glass had been hotter the water would have divided it more finely and would have extracted a greater quantity of gall. On account of this I could not doubt that the gall was the cause of the bubbles.

The pot in which I had made these three experiments contained about two hundred and fifty pounds of glass; one part fixed alkali salt extracted from tobacco ash and one and a half parts of white sand comprised the batch.

We shall now add a few observations which seem sufficient proof of the truth of what we believe has just been established. The test dips drawn from the pots always contain fewer bubbles as the fumes approach their cessation. At the beginning of founding it is not rare to find tests drops that are hollow and that their cavities are a quarter to a third full of pure gall. There are many fewer bubbles in the glass, other things being equal, the greater the heat and the longer the founding time. In small ordinary glass works the heat is feeble and the glass is worked as soon as the fumes have ceased; their glass is thus full of bubbles. At the moment when one stirs the glass in a large pot, the surface appears to be nothing but bubbles, yet if one takes a sample dip from a depth of two inches, there are far fewer: this effect is without doubt due to the vapour which has been released and risen from the bottom as a result of the movement of the instrument that has been used. The more that the glass produced by a well-proportioned batch has been cast very hot into cold water, the fewer bubbles does it contain, having lost part of its gall to the water on each occasion.

The custom of casting glass into water is very ancient but the operation was not undertaken with a view to preventing bubbles; *it is* (says Neri) *so that the gall can be separated, because this salt is harmful to crystal,* which it renders obscure and cloudy, and which the crystal forces to the surface during working.

However certain it appeared to me that gall turned to vapour was the cause of bubbles, one thing remained before I would be satisfied, namely to see a reasonable quantity of glass purged of gall in which I could find no bubble. For this purpose I made a very soft glass. It was melted four times and cast into cold water each time: on the fifth melting I stirred it well with a very clean iron rod; no bubbles at all appeared in the dips that I took from it. I then put gall into the pot in several stages and mixed it well with the glass: bubbles reappeared in the test dips.

One rigorous test remained to be made with the glass purged of gall so far as it could be undertaken: I believed that Prince Rupert's drops would be sensitive to the very smallest amount of gall vapour in the glass that, forced towards the centre of the drop by the chilling of the surface, it would produce one or more bubbles. If I obtained Prince Rupert's drops without bubbles I would thereby have carried my discovery to the highest degree of certainty and I would have decided a question which had occupied the greatest physicists of Europe; to know whether the bubbles which they had always contained were essential to or inseparable from them. I had many drops of our glass cast, in the accustomed manner, into a bucket of cold water; the drops that I found in the bottom of the bucket contained no bubbles, broke with vigour and were reduced to an enormous number of particles when the tails were broken. I had the honour to demonstrate this last year to the abbé Nollet and M. le Camus. Every time that I have repeated the experiment it has had the same success: I did not forget to mix gall with the glass and the Prince Rupert's drops formed from that did contain bubbles.

I have observed repeatedly that: 1) one produces a much smaller number of successful drops with glass purged of gall than with any other, very probably for the reason that the glass in the centre yields less than the vapour to the violent contraction that quenching causes in the outer layers: 2) the hotter they are, the fewer drops fail: 3) the drops produce a greater effect as the water into which they are poured is colder.

It is, I believe, clearly demonstrated that bubbles only occur by chance in Prince Rupert's drops: it seems to me no less certain that those who have attempted to explain the peculiar property of the drops by the action of air or, like Rohault and Polinière, by the flow and action of rarefied materials invented at will, have been very far from the truth; it is to the *abbé* Nollet that the honour of giving the true explanation must be given. One could not see anything more satisfying or conforming better with experience than what he says on this subject towards the end of volume four of his *Lessons in Experimental Physics*.

When one carefully examines the shape and colours of a large Prince Rupert's drop, it seems to me that one cannot doubt that the glass has been cooled by layers and that the inner ones, on cooling, have been obliged to contract towards the outer ones which have already been cooled.

Particles of dust, a fragment of lint or wood, a drop of water or anything else that can produce vapour on burning, can produce bubbles but these particular sources are of little importance and, with only a little care, the workmen can avoid their effects. [However] this is not true of the general cause that we believe we have demonstrated, which has its source in the batch and is not easy to avoid.

In glass houses where the heating is stopped as soon as the glass is judged to have been founded sufficiently, the furnace is stopped closely and the glass is then worked only when it has acquired, by a slight diminution of the heat, an appropriate consistency [viscosity]. It is believed that the glass has thereby been allowed time and opportunity to expel the air as it settles or densifies. If there are no bubbles in the glass, or very few, this is not because the air has been dissipated; rather because the gall vapour has not been collected together but has been equally dispersed through the whole mass of glass. Experience has shown only too clearly that glasses in which bubbles are rare contain a lot of gall concealed within their substance: if exposed to a damp heat *they transport it towards the surface and sweat it out*. Nearly all our crystal glass falls within this category. If a moderate sized piece of this soft crystal is made red hot in a furnace and the fire needed to keep it very soft is suddenly removed, a great number of bubbles are formed. This phenomenon was previously inexplicable but it is today easy to give the reason. I will have occasion, in another memoir, to show why the most common glasses do not expel any gall although they contain scarcely any less than the better quality glasses.

Without any doubt the best way to prevent bubbles is to purge the glass of gall: this may be done by casting the very hot glass into water, by stirring it many times with a stick of green wood, by the shock of a bullet, by mixing it with an iron ladle, by introducing into it volatile materials such as arsenic, antimony, &c., by using the most violent and long continued heat, and, above all, by using well proportioned batches. The separation of the gall as a layer floating on the melt is of no importance in determining whether the bubbles are removed or not. It may be expected that I will here give the most advantageous batch compositions but such detail would lead us too far and will much better find its place in other memoirs. We will also defer to another occasion the reservations that should be noted in relation to what has been written about this by Merrett and Kunckel. One says that *a pot containing two hundred pounds of materials of the best quality will give up to fifty pounds of gall*; the other *that all the salts extracted from the ashes of vegetables are of the same nature*.

Before concluding this memoir it may not be without point to say that gall is no less the cause of bubbles in enamels than in glasses: this could be confirmed by taking the trouble to apply to the enamel the observations and some of the experiments that we have made in relation to bubbles in glasses. The means that we have indicated for preventing them in the one apply equally well to the other.

We have no change to make in this memoir; twenty years additional experience have confirmed what it contains. In 1758 the art of glass making was so little understood and [earlier] authors who had treated it enjoyed so high a reputation that my memoir on the perfection of glass making would not have been so favourably received in 1760 if I had not taken the precaution of preparing people's minds with this one.

A few years ago a provincial Academy proposed, for a prize, the same subject, *The discovery of the cause of bubbles in glass*. This worthy body presumably had no knowledge of my memoir.

MEMOIR ON THE NATURE OF THE ELECTRICAL FLUID WHICH PROVES THAT GLASS ITSELF IS NOT ELECTRIC

Read on 17 December 1762
and printed in the second volume
of the Academy of Dijon

THE discovery of electricity appeared so curious that physicists have been occupied with it, especially in the New World. There is no other material to which so many experiments or, perhaps, so much wisdom have been devoted. The *abbé* Nollet, in particular, seems to have exhausted the subject.

The electrical phenomena have, so to say, concentrated the attention of savants who have been satisfied with giving their conjectures on the electrical fluid. It is the element of fire united with certain parts of the electrifying body or that of the body being electrified; or the medium by which it is transferred; a material of the same nature as animal spirits, similar to that of thunder, &c. These ideas have not been confirmed by experiment. I venture to hope that some more precise and satisfying facts can be found in this memoir.

Researches have been pushed very far, not only establishing which materials possess the electrical property but even to put each of them in order; glass is recognized by all physicists as the most electrical body. The most evident appearances agree with this opinion but it is no less true that glass itself is not electric; it is only by another different material that it can normally be charged.

Some years ago I made two discoveries that, to be precise, are in fact one. The desire to repeat the experiments did not allow me to bring it to light. Towards the end of 1756 I contented myself with announcing it in a letter to the *abbé* Nollet which he was so good as to read to the Academy of Sciences in Paris.

In July 1756 I had the honour of seeing the *abbé* at Saint-Gobain.

He asked me to make him some glass tubes, which a workman did at once. The *abbé* was surprised to to see that none of these tubes showed any sign of electricity even after the most vigorous and long continued rubbing. Much thought was devoted to this phenomenon and, suspecting that the state of the atmosphere might be responsible, the finding was scrupulously re-examined in subsequent days. It was confirmed that these tubes, rubbed in dry or humid weather, indoors or in the open air, never produced any sparks. The possibility that this might be due to the particular nature of plate glass was rejected. Some people might think that this glass exposed for a long time to a very violent heat, had become so compact that friction found it difficult to set the parts in motion and that the sparks could not escape. This conjecture appeared sufficiently reasonable for the *abbé* to ask me to put it to the test. I therefore made a trial to satisfy his wishes.

As soon as the batch had melted and and the gall been dissipated, I had some tubes made. These were rubbed with care, because of their fragility, and they gave signs of electricity. This was repeated many times with the same success. These experiments thus appeared very favourable to the conjecture made above. However, I thought it important to push my researches further. I had tubes made at all stages of melting and I observed that electricity diminished in proportion as the refining of the melt approached completion. In that state the glass had no electricity. These experiments were varied as much as possible but always gave me the same results. I thus came to accept the conjecture as a fact but was persuaded that one cannot be too cautious in making pronouncements, even after making an experiment, so I reflected that, whilst losing its electrical property during founding, the glass also lost its gall and its gross colouring matter, the latter being provided by the manganese. I therefore examined whether the dissipation of one of these materials, or of both together, would provide the cause of the phenomenon.

For this purpose I had very white gall cast onto well-refined glass not showing any signs of electricity. After mixing that as well as possible I had tubes blown and they showed no signs of electricity. New experiments with gall showed no different results.

It thus remained to discover whether the colouring principle contributed to the electricity of the glass. I did not waste a moment to examine that. I mixed into glass well-refined and non-electric a sufficiently large quantity of manganese to make the glass dark red. The tubes that were made did give electric sparks. This experiment, repeated with care began to change my ideas about the nature of the electrical fluid. Far from being discouraged, I worked with renewed zeal to reveal most clearly the cause that I sought.

I mixed into non-electric glass materials that contained the largest amounts of colouring principle, those most appropriate to reduce metallic oxides and able to produce sulphur from Glauber's admirable salt or sulphated tartar by Stahl's method, zaffre, ordinary powdered carbon, chimney soot, resins, Spanish wax, animal materials reduced to very black carbon, &c., &c. The tubes that were blown from glass to which colouring principle had been added as one or more of these materials were very electric.

These wide-ranging experiments appeared to prove that the electricity of the glass was due to the colouring principle but a doubt arose in my mind that I sought to dissipate. Might it not be that some other principle in the materials that we use, other than the colorant, that makes the glass electric?. To convince myself, I dissipated the colouring principle as completely as possible from the materials mentioned by means of a clear flame for a long time. Deprived of this principle they produced no change in any non-electric glass with which I mixed them. But is the colouring principle itself the electrical fluid or is the latter developed by it in the other materials employed or in the glass with which they are mixed?

Although this doubt did not seem to be as well-founded as the previous one, I made the following experiments to convince myself. having carefully mixed equal parts of unfritted soda from Alicante and sand, I filled the mixture into a pot. This batch gave me a very black glass, almost opaque and very electric, even after being kept in the furnace for double the time used for normal plate glass batches. I then mixed into this black glass very white gall and repeated that many times until the colorant was entirely dissipated. This procedure made the glass pass through all the gradations of colour from black and all shades of yellow to the least disagreeable clear green. At that stage tubes were drawn and I was convinced that the glass had lost its electricity as well as its colouring principle. I then mixed in, both separately and together, chimney soot, powdered carbon, resins reduced to carbon, &c., which made the glass dark yellow and almost as electric as the black glass. I also added enough zaffre to make the glass deep blue and the tubes made with that were strongly electric. I then had no further doubts that the colouring principle itself is the electrical fluid and that tubes made during these experiments that showed no electric effect did so because they had been deprived of all or most of their colouring principle. Hence the glass itself is not electric.

However, a difficulty which appears to contradict our experience now presents itself: passably white [pale tinted] glass can very electric. I had so little doubt of this fact that I sought to make a tube that united these two apparent extremes. The glass of this tube, which appeared sufficiently white contained more colouring principle than the yellow and blue glasses mentioned above. To be convinced one must note the following:

1. That the bubbles of which it is full and make it appear milky, show that the glass is insufficiently founded and refined, that it contains a lot of gall. But the neutral salts are much less tinted by the colouring principle than is the glass. Tartar and sugar appear white but burn and become black when heated and can detonate nitre.

2. When the three prime colours, yellow, blue, and red, are present in a glass in appropriate proportions and the latter dominates to the least possible extent, an agreeable white is produced. I flatter myself to have shown that in my essay which won the prize offered by the Royal Academy of Sciences.

3. That the glass of this tube is very obviously tinted by the red of manganese.

Many people will ask me what I understand by the *colouring principle*? I cannot reply, like the ancient philosophers, that this exists only in our soul, nor with the celebrated Newton that it is a property exclusive to the rays of light, but with the great Stahl and the knowledgable Pott that it is what all the chemists call the inflammable principle, *phlogiston*. If any doubt remains in this regard, the experiments of which we have given an account seem to me very appropriate to dispel it. The reasons for not regarding phlogiston and the colouring principle as one and the same thing are not clear: it reduces metal oxides, it makes sulphur from sulphuric acid it detonates saltpetre.

I do not believe that non-electric glass is completely devoid of phlogiston. What body in the universe could be in such a state? It is probable that for glass to give signs of electricity the phlogiston should not be too attenuated or should be there in a certain proportion. What is incontestable is that more of it makes the glass more electric.

The following long footnote, presumably written in 1780, is inserted here:
It does not seem to me that the new discoveries of the knowledgable Sage, some of which have been confirmed by Scheele, allow us to doubt that every type of glass contains an acid as a constituent: this acid is definitely not sulphuric, nitric or hydrochloric but *acidum pingue*, animal acid or vegetable acid in its greatest degree of purity; the acid that is a constituent of calcareous earths, alkali salts both fixed and volatile, in fusible spars, basalts, &c. This acid, the densest of all acids, is the most fixed and thus made active by fire is probably the only vitrifying principle. This acid has a greater affinity for phlogiston than any other acid and thus can strip them of it and by this means become volatile and extremely elastic. Combined with a certain quantity of inflammable principle it forms a true *sulphur* known by the name *phosphorus*. So combined the phlogiston being in a smaller quantity than is needed to solidify it is is made luminous by the slightest movement that it receives from heat or friction and it expands giving it a smell of garlic. These principles being put forward, the truth of which I cannot doubt because Sage has demonstrated his experiments to me with the good nature that one expects of this true savant and which make him particularly notable among my acquaintance. Hence it was

easy for me to conceive the interaction of the electric fluid and glass. The motion communicated to the phlogiston, and assisted by the elasticity of the glass itself permits the conclusion that any obscurities concerning the primary electric phenomena can be removed by a more exact knowledge of the nature of glass. In glass where there is the least possible quantity of colouring principle or phlogiston, friction can impart a violent motion to the acid, making its integral parts rotate on their axes without being displaced, without giving any other sign of electricity. The contrary should occur in a glass charged with inflammable principle and the phenomena should be much more detectable as there is much more phlogiston; nothing agrees more with experience. The glass becomes more electric as it becomes more coloured and *vice versa*. Thus this acid seems to me the vehicle necessary for the electric fluid, as it is for the light [luminescence] in all possible cases. Too many relations between phosphorescence and electricity have been observed not to realize their perfect identity.

* * * * * * * * * *

Certain observations should have led to the suspicion that the electric fluid is nothing other than the colouring principle, phlogiston. The experience of all those connected with electricity is that blackest bottles are the most electric. The *abbé* Nollet noted that glass was made more electric by blue enamel with which he had coloured it. There is no one who has not observed that the most electric materials are those most charged with phlogiston, such as amber, sealing wax, &c. It has even been discovered that wood can be impregnated with condensed phlogiston to supplement, in certain cases, electric materials, for example supports or cakes of resin smelling of garlic, of burnt arsenic, of the dissolution of iron to produce electricity. These things would, it seems to me, have raised suspicions. It appears very probable to me that phlogiston is no less the principle of odours as well as colours. Of the three blacks produced by the primitive colours when condensed, the black of the yellow has always given me the most electricity, presumably for the reason that it requires a greater amount of phlogiston.

I believe that glass is the best material to produce electricity by friction and the best for making experiments with electricity because the phlogiston is intimately united with it and because the body is perfectly elastic.

I think that we are now able to appreciate the mystery raised by some master glass makers of the best glass to use for electrical experiments. For this reason searches have often, in the past, been made afar for what lies at one's door. A master glass maker of repute in 1757 sent me, as a great mark of friendship, what he called the secret of making the best glass for electricity and begged me not to communicate it to anyone else. He prescribed, among other things of little importance in his recipe, that the only flux to be used in his batch must be ash of oak. In thanking him I mentioned that he did not have the secret of making electric glass, that the blackest was the best and that equal parts of yellow clayey sand and raw soda would provide the best and least inconvenient. The arts swarm with prejudices!

It does not seem to me appropriate to go further into the usefulness of my discovery which I would like to extend considerably. Experimental physics should gain in proportion as it is possible to simplify the principles. savants who who start from proved points of departure will surely be those who provide the most certain explanations of electrical phenomena. They should perhaps have the pleasure of envisaging electricity from new points of view and they may be able to discover the analogies between the electrical fluid and animal spirits, meteors, &c.

MEMOIR ON THE NATURE AND CAUSE OF THE DIFFERENT TYPES OF SMEARS IN GLASS

printed in the eighth volume
of the Savants Étrangers,
presented on 7 December 1765

IN MY memoir on the cause of bubbles in glass I undertook to discuss the other effects of gall. The desire to carry my researches further and produce more evidence has not permitted me to fulfil this pledge before now.

It was not possible for me to develop further the effects of gall in my Memoir on glass making; I had to be content to indicate them.

What remains for me to do is much too extensive to be conveniently treated in a memoir; I shall restrict myself today to the *smears* or opalescence that gall produces in glass.[a]

My work will not be full of citations: I do not know any other author who has discussed this subject. *Neri*[b] said, it is true, that alkali salt (*gall* which he regarded as a superfluous alkali) makes crystal translucent and cloudy, but this idea is so little developed that it did not seem to merit any attention.

Smears in glass are hardly known in glass houses other than those working in crystal or colourless glass. There are many reasons why they are very rarely observed in ordinary glass works: 1) because these works never use pure fixed alkali; 2) that they use a much smaller quantity of fixed alkali and hence of gall in their batches than is used in quality glass; 3) because

---

a. See p.41 of my Memoir on the *perfection of glass making*
b. A. Neri: see p.39 of his *Art of Glass*, 4º French translation

the lime contained in soda or ordinary ash gives a greater fluidity to the glass and this favours evaporation of the gall; 4) also because the colouring agent, the *phlogiston*, which the ashes of *kali* and other plants contain in abundance, makes the gall volatilize during melting.

Glass houses making fine quality or miscellaneous decorative ware are the most likely to find their glass too fat. The low prices that circumstances have produced in this branch of the industry, force them to be more economical with fixed alkali salt than plate or Bohemian style flat glass works, &c. and we will, in a moment, make this clearer.

Sometimes, at the moment when it is desired to work the glass, it is found unsuitable for making saleable ware. It is semi-transparent, the colour of milk and lacking strength; at the first glance it might be taken for porcelain or chalk glass: this is the first degree of opalescence and the most harmful. This type of opalescence is, like the others, variable; the milky colour may have either a green or a blue cast. Several very similar things are experienced in ordinary glass houses, particularly in bottle making: their glass takes on more or less the colour of milk, sometimes in the pot, sometimes when reheating the article. There are, I think, few people who have not seen bottles suffering from this defect, having areas showing whitish or bluish streaks. In the jargon of the trade this is called *veiled**. Whatever resemblance this may have to the opalescence we have just described, it must not be confused with it: they have very difference causes. Bottle glass only develops the opalescence because a certain part of its lime has not been vitrified or because this glass has decomposed and been converted into porcelain.

On other occasions, at the beginning of or during working, one sees throughout the substance of the cooled glass a whitish cloudiness and very large numbers of extremely small white specks; the glass is off-white, less brilliant, and less transparent than usual for good glass and breaks more easily during annealing or in the warehouse: this is the second degree of opalescence.

---

* The French term is *chapeau*.

There are drinking glasses and other pieces which appear to have been badly rinsed out; one believes that one can see traces of greasy finger prints or a slight cloudiness of a colour different from that of the glass: domestic servants then become the victims of the incompetence or negligence of the manufacturer. This is the third degree of opalescence.

When one examines blown flat glass that has been made with soda, a smokiness of varying density, violet or tending to violet, may be seen: this defect does not extend over the whole sheet but affects one or more separate areas. This is the fourth degree of opalescence.

I have seen, more than once, blown plate which appeared contaminated throughout its substance with a very great number of tiny knots and which, however well it has been polished, always appears rough. This is our fifth and last degree of opalescence.

It is not correct to call all these types of defects smears but they are given this description because glass which is thus infected, does when penetrated by moisture, genuinely become rather greasy to the touch.

Opalescent glass is more easily broken, as we have already observed, but it is also much more susceptible to attack by moisture and more difficult for it to develop the red colour due to manganese as it becomes more opalescent. These two phenomena led me naturally to the cause of the opalescence. It is clear that the gall, having been dried out in the pot and on evaporation carried away with it the colouring agent in the manganese, attracts moisture from the air when the glass is cold.

The most direct paths are rarely the first that one tries. That which I took, far from taking me to the objective that I sought, took me further afield. I questioned the directors, the master glass founders and the principal workmen in different glass works; nearly all concurred in attributing the first, second, and fifth degrees of opalescence to a chill of the glass in the pot, that is to say a decrease in the heat, the third and fourth degrees to smoke from the wood, to which, because of the imprudence of the *teaser*, the article was exposed when it was being reheated in the working hole.

It is true that, in glass houses making miscellaneous ware, the glass becomes more or less opalescent [it devitrifies] every time that the teaser neglects his furnace.

Whenever one puts hot glass in contact with a piece of wood the smoke makes a mark on the glass which cannot be removed. These two facts, however conclusive they appeared to be to the people whom I consulted, left some doubts in my mind. I have seen glass which never became opalescent in a feeble heat. At the same degree of heat opalescent glass had always seemed to me more viscous, harder [to work] in the pot, than glass free from the defect. For a long time I had known that glass became more subject to this defect as one diminished the quantity of fixed alkali used in the batch and as this fixed alkali was better purified of its colouring principle. The mark that wood smoke made on the glass did not appear to me to be of the same colour as opalescence of the third and fourth degrees; this mark could be removed by grinding away a very thin layer of the glass but the opalescence of the third and fourth degrees occurs through the whole thickness of the glass: furthermore opalescence of the fourth degree affects only one or more parts of the plate; if it were caused by smoke it would affect the whole surface. This state of uncertainty about a phenomenon which I often had occasion to observe was too distasteful for me not to put in train everything that seemed likely to elucidate it. I thus considered the matter from a new point of view but, before giving an account of my experiments and demonstrating what I found, it is necessary to observe that the proportion of fixed alkali used in the batch for colourless glass should be related to the heat of the furnace. To make this matter easier to understand, suppose that glass making furnaces give many specific degrees of heat, and that five degrees of heat [are needed] to melt and effectively refine a batch of two parts fixed alkali and three parts sand. With the same batch, one will have the first degree of opalescence if only the third degree of heat is achieved; with only the second degree of heat the materials will not melt properly, &c. If batch of one part fixed alkali to two of sand is taken to the fifth degree of heat, the second degree of opalescence will be found; at the fourth degree of heat the first degree of opalescence will occur and at the third degree of heat the batch will not melt. It can be seen that the quantity of fixed alkali in the batch varies according to whether it contains more or less gall. One of the great difficulties of the practical arts is that one nearly always has to take many different factors into account.

The examination that I made of glass as it begins to melt gave me my first insight into the cause of gall. In this state the glass is of poor transparency, more or less milky, easily broken and very susceptible to attack by moisture, it never shows the colour of manganese, it is opalescent. If the heat is sufficient, the batch made in appropriate proportions, and the fixed alkali salt not too rich in gall, the opalescence will disappear by degrees.

What does the glass lose during founding or refining? Innumerable experiments have convinced me that the glass loses only its colouring agent and the gall. I could thus attribute opalescence in the glass taken at the beginning of refining only to one or the other of these two causes, or both together.

With a view to determining the truth I mixed powdered wood charcoal, a colouring agent, into the glass. Far from increasing the opalescence, this inflammable material dissipated it. However natural it might seem to conclude that the gall was the source that I sought, I wished to confirm it by a new experiment. I had some very white gall mixed into the glass and the opalescence reappeared; I could only dissipate it by means of materials containing a great deal of phlogiston.

Although I no longer had any doubts about the cause of the opalescence, I thought it desirable to make further researches to establish the strongest possible evidence for my discovery. Opalescent glass examined under a hand lens is seen to contain innumerable white specks exactly like gall seen under the same lens.

I had very opalescent glass cast into water: when the water was evaporated a residue of gall remained and when remelted the glass was much less opalescent. A very abundant whitish vapour is produced when the opalescence is removed by powdered carbon or by any other violent means. If this vapour is condensed it is found to be nothing other than gall. If gall is mixed with a glass which has been freed of opalescence by phlogiston, it again becomes cloudy. I had a very opalescent glass mixed with fixed alkali salt, purified as much as possible, and the defect disappeared; when gall was well mixed with this glass the opalescence reappeared for only a very short time. This phenomenon showed the reason why works making plate glass, &c., are less likely to see this defect in their glass than those making miscellaneous ware. The addition of fixed alkali salt makes the glass less viscous and consequently favours the evaporation of the gall.

I believe that the experiments that have just been described show clear evidence that gall is the natural cause of the first, second and third degrees of opalescence but do they also prove that it is the cause of the fourth and fifth degrees? I think not and must admit that these two degrees of opalescence tortured my mind for a long time.

Why is the third degree peculiar to blown plate, for which the glass was melted with soda? How can gall produce the knots which characterize the fifth degree? Whatever difficulties these questions posed for me, I neglected nothing that might help to resolve them. To make clearer what has been said about the fourth and fifth degrees of opalescence, we will consider them separately.

If the parts of the plate glass affected by the fourth degree are examined with a hand lens, the cloudiness can be confirmed to be due to a collection of very small white specks similar to gall; the bluish colour of the glass and the red colour of the manganese give this cloud a violet tint.

I have also frequently observed that the parts of the plate so affected are more susceptible to attack by moisture, becoming fogged or weathered and developing *partridge eyes* in the warehouse more easily than the other parts. We have already seen above that one of the characteristics of smears is to make the glass more susceptible to moisture.

However, I have never been able to find this type of defect in glass made with the best Alicante soda, used in appropriate proportions, and subjected to a violent and long continued heat.

It is generally known, as I have often experienced, that soda from Carthage contains half as much gall again as that of Alicante and that it usually produces the fourth degree of opalescence.

All these observations appear to prove sufficiently clearly that gall is no less the cause of the fourth degree of opalescence than of the first three, but they do not tell us why this defect is specific to blown plate or why it affects only one or more areas of the whole plate. This peculiarity was the subject of further researches by me.

Blown plate is annealed in two different ways: in flattening furnaces and annealing kilns. The flattening furnaces are very long and the crown is very high on the side opposite to that where the fire is. The plates are flattened in the end where the crown is low, then the plates, flattened on the hearth of the furnace, are pushed towards the other end where they are stood on edge and held by iron bars which fit horizontally across the furnace at different heights. The sheets are judged ready to put on the stack when they sufficiently cool not to sag [under their own weight].

The annealing kilns are furnaces twenty five to thirty feet long and eight to ten feet wide. This type of furnace is heated from both ends. There is a special furnace for flattening the glass sheets from which they are carried to the annealing kiln on a large iron paddle where they are placed side by side. Before putting the sheets into the kilns they must be heated just to the softening point of the glass and they are kept hot for twelve hours as other sheets are successively put in place. The necessity for this digression will be seen in a moment.

However much care that I have taken, I have never been able to discover any opalescence of the fourth degree in glass annealed in flattening furnaces; all those that I have seen with this defect have been annealed in kilns.

The gall contained in soda, that is to say Glauber's salt and sea salt, is more volatile than the gall from potash, sulphated tartar. This gall receives an increased degree of volatility from the phlogiston which is supplied to it very abundantly by the lime in the soda. It is sufficient for the glass to be hot enough to scorch wood for this gall to be evaporated to some extent. We can consider vaporized gall to be a fluid which will escape by whatever means offers least resistance. However, it is undoubtedly true that the thickest parts of the plate retain their heat much longer than the thinnest parts and that this heat is more intense in the middle of the glass than at its surface, in this thickest part. The gall vapour can thus flow towards the thinnest parts of the sheet and condense there when the fire is damped down or as a result of cooling. In other words I have always found that the opalescence only occurs in the thickest parts of the plate glass.

Although this explanation squares well with the principles, it seemed only a conjecture to me. I was eager to confirm it by experiment: I mixed a small quantity of gall with some glass which had been completely freed from all trace of opalescence. The worst workman was then chosen to make two pieces of plate; he made them, as I had expected and desired, of very unequal thickness. They were annealed in a kiln. The thickest parts of these two pieces were affected by opalescence.

The fifth degree of opalescence need not detain us so long. One usually only detects it near the end of working, that is to say when the glass in the pot has been worked for five or six hours continuously, when the glass is too low in the pot and the heat of the furnace too feeble for the gall, which continually rises to the surface of the glass, to be entirely dissipated. It then forms as a film of gall on top of the glass. The workman plunges his large blow pipe three our four inches into the glass and rotates it a few times for the glass to adhere to it. Within a few minutes he does this operation three times to gather as much glass as is necessary. Whoever can properly understand this manoeuvre will see how the gall can appear distributed throughout the finished glass as small knots. These knots are only very small particles of gall; with a hand lens one can see distinctly the gall that comprises them. I confirmed this in a more decisive manner.

Having allowed the heat of the furnace to die down I threw a small quantity of gall onto the surface of some glass which had no appearance of opalescence: opalescence of the fifth degree was very obvious in the plate glass that was then made from it.

Our researches would only be complete, our curiosity satisfied, if they led to the discovery of the cause of smears. We only say this because they showed us means as simple as efficacious to avoid one of the greatest

On Glass Making

enemies dreaded by master glass makers.

In glass houses, without knowing anything about the cause of the defect, remedies have been found; each has its own which kept a great secret. The majority of these remedies are palliatives or specific cures: let us cast an eye over several of them. When master glass makers perceive that their glass is opalescent, or tends towards opalescence, some mix into it, in many doses, manganese either powdered or in lumps, arsenic, powdered carbon, wood bark especially that of aspen, willow or birch, &c.; others stir their glass with stick of green aspen, lime, ash, hazel, birch, &c. 7 to 8 inches in circumference.

These methods only partly remove the opalescence, it reappears to some extent a short time after the operation, or during working, and it is very rare for the glass not to retain some traces of the treatment. The reason for the inadequacy of these methods is obvious; with however much care they are used, they never act equally on all parts of the glass contained in the pot. In addition they have three inconveniences which should lead to proscription: 1) they occupy precious time; 2) they may lose a lot of glass by making it boil up and spill over the rim of the pot; 3) the violent agitation of the glass frequently gives bad refining leaving the melt full of seed and the dissipation of the colour due to manganese is incomplete. There are some glass works that employ much more effective methods which could be regarded as specific remedies: some use only red potash, not white, in their batches, the phlogiston which is abundant in the former makes the gall volatilize and certainly prevents opalescence; others carefully mix a certain proportion of powdered wood charcoal into their batch made with white potash, the phlogiston in the charcoal necessarily producing the same effect as that of the red potash.

There are some very intelligent and very experienced men who can successfully apply both these methods: If the red potash is lean, hardly greasy, too rich in gall, the batch will produce opalescence. If the red potash is adulterated, which is common, if it is contaminated with foreign materials such as soot, sap, manure, or clay, the batch will be difficult to melt and the glass a disagreeable yellow green; it then takes the colour of manganese badly. If insufficient powdered charcoal is mixed into the batch it will be more or less opalescent; if too much is used, the glass will be a very unsatisfactory yellow-green colour. It is very difficult to find the correct proportion of charcoal because it must be related to the purity of the raw materials and to the quantity of gall in the potash. Besides that, it is very rare that the powdered charcoal does not make the melt froth so that some overflows the pot. Experience will prove that these are not ill-founded fears.

Since the means that glass works currently use are insufficient, or of uncertain effect, or subject to considerable disadvantages, it is natural that we should neglect nothing to discover better ones which would both meet the views of the master glass makers and contribute to the public good. It was to this that we applied ourselves with renewed ardour: we believe that our researches have not been without success.

The proportions that we have given four the third batch in our *Memoir on the Perfection of Glass Making* would be an infallible means of preventing opalescence but it would be too expensive for the master glass maker producing miscellaneous wares: if he used more than a third of fixed alkali in his batches he could not compete with others.

Being sure that gall produces opalescence, I sought to make the effect impossible by destroying the cause. The different quantities of water that fixed alkali salt and gall of potash glass, potassium sulphate, require for their dissolution, provided me with a means both simple and certain.

I placed a woollen blanket folded in four over a barrel then I placed on this type of filter 200 pounds of well calcined and well crushed potash, then I poured over the potash sufficient hot water to fill or almost fill the barrel. After this operation I found on the filter a considerable quantity of a white salt, in parallelepiped crystals which gave no effervescence with acids, in other words the true sulphated tartar [K_2SO_4]. After the third dissolution filtration and evaporation no or very little gall remained mixed with the fixed alkali salt. The dissolution may be done using cold water and two filtrations suffice in this case but the operation is very tedious and it requires a much greater quantity of water. One part of fixed alkali salt, thus purified, mixed with two parts of sand melts rapidly and the glass produced never shows any sign of opalescence. A new proof that gall is the cause of opalescence.

This discovery is not only useful to the art of glass making but it also offers an important resource to dyeing and also to medicine: we may, perhaps, one day develop these ideas.

However simple may be the means that we have just indicated for preventing opalescence, it probably will not be to the taste of all master glass makers; we can recommend to them another which is just as efficacious, which requires much less apparatus and which is just as applicable to soda and alkali nitrate as to potash. Lime calcined in air or quenched in water from a high temperature considerably increases the fluidity of the glass. This lime is very suitable for making the gall volatilize: one can thus use lime as an infallible means of preventing opalescence. A very large number of experiments have shown me that the best proportion of lime that can be used is one twelfth of the batch; that is to say that for 100 pounds of good white potash and 175 pounds of sand one should use 25 pounds of lime.

I am not ignorant that the use of lime is known in the German glass houses and some in Alsace, but I do not believe that any of them use it to prevent opalescence or that they have any reliable rule for the proportion to be used in the batch.

OBSERVATIONS ON THE ART OF MAKING FAIENCE
Read to the Academy of Dijon and printed in the first volume of the same Academy [1769]

THE manufacture of faience is an important part of the art of glass making. It has been no less neglected than the others. It even seems that the chemists have avoided talking about it. I know that Kunckel took pains to give a few recipes for glazes and paints for faience;[a] but I doubt that they have been of much use. As long as the arts have only recipes for theory they will remain very far from perfection. The making of faience is an unequivocal proof of that.

In France there are only two manufacturers of ordinary faience that enjoy a reputation, Moustier and Rouen, and their merits owe more to local circumstances than to any principles on which their work is established.

The faience of Saint-Cenys in Picardy was long ago very sought after. It has now, with good reason, fallen into disrepute but is beginning to re-establish its reputation. I know of entrepreneurs who abandoned their manufactures because they could not make their enamels brilliant; others who could only make their white enamel take on parts of the biscuit; others because they could not prevent spalling, &c.

The faience works of L'Isle [Lille] in Flanders, of Saint-Cenys, of Lyon, of Nantes, of Rouen, &c. obtain their sand from Nevers although they have the whitest at their gates. One sees a lot of faience that shatters, of which the enamel crazes at the slightest heat; one sees very little that is not infected by flaking and even less that does not suffer from specks or marks left by greasy fingers [French: *écoussages*], or eggshell defects in the glaze [French: *coque d'œuf*], &c. Who does not feel that this art is given

---

a. See the Art of Glassmaking, quarto, pp368 et seq; 407 et seq.

over to blind routine! I do not propose to give a complete treatise on making faience nor to describe in order all its operations; such an objective is beyond my abilities. I will confine myself to a few observations that I have had occasion to make in a works making good fine and ordinary faience that one of my near relations established some years ago. It is for the Academy to judge how far they can contribute to progress in the art.

Enamel of the best and whitest quality, the most brilliant colours and of the most perfect match, the best constructed furnaces, the most skilled and experienced workmen, would be useless to a manufacturer of faience if the clay were not of a good nature, properly blended and well prepared. Incompetence and negligence in this regard can only bring him sorrow. He will inevitably be ruined by breakages during drying and in the furnace, or by distortion of the ware, or by streaks (matt enamel), by flaking, &c.

Every one knows that in the majority of faience works, only common earths green or blue potters' clays, reddish, yellowish, grey, or brown earthenware clays are used. (I do not intend to speak of pipe clays here nor of the English type [of faience] which only differs in the glaze and which, until now, has been so badly imitated). Those two types are not our object.

The works in Paris employ for their common faience, the greenish potter's clay of Belleville, the yellowish earthenware clay of Charonne, and the whitish marl of the sides of the Picpus; they add to their brown, or fire clay, potter's clay from Arceuil. At Thionville, at Aprey, &c. three types of clay very similar to those used in Paris are used. At Nevers they use only two types of clay for faience, a fat yellowish clay and white marl. I believe that there are few faience works sufficiently well situated to be able to use only one clay.

Blue, green, or grey pottery clay seems to me to be only pure clay charged with ferruginous substance, a small quantity of lime more or less coarse, a little sulphuric acid,[a] and sometimes a little very fine sand.

Reddish, yellowish or brown clay, or common brick clay only differs

---

a. See p. 33, vol. 1 of Wallerius's Mineralogy

from potter's clay in the ferruginous base being more abundant. That of Nevers holds the middle ground between the other two and it may be used to make excellent hard bricks. No one can ignore the fact that marl is a calcareous earth; however, that which is used in faience works is mixed with a little clay and very commonly with a little iron-rich matter.

For what reason is one obliged to use many types of clay? Wares made with potter's clay alone would take too long to dry, crack and distort in both drying and firing, would be insupportably heavy, and would be rejected. They need an intermediary which can prevent too great a shrinkage, make them less dense and does not allow the enamel to attack the body. Red and similar fireclay is never less than able to confirm these views; it is to be feared that it would suffer from almost the same inconveniences and that the ware would be more inclined to melt. Marl provides what is desired; it decreases shrinkage to an acceptable level, allows the water to escape easily [during drying] without straining the ware and, other things being equal, produces a white body, a glossier and more brilliant glaze because with its aid, the other clays being less inclined to melt, cannot combine too intimately with the enamel or, if one wishes, it gives to the enamel what the other earths make it lose.

It is known that a glass approaches more closely to a good white enamel if an excessively large quantity of lime-rich earth has been added; the well purified calcareous earth produces in the enamel almost the same effects as tin oxide. Those who might conclude from what we have just said that reddish fireclay is useless seem to me to mislead themselves. Wares made solely with potters clay, with a suitable proportion of marl for the white, would not be sufficiently solid and would spall unless one did not subject them to a greater heat than that used in the ordinary faience works. It is the reddish fireclay that because of its ferruginous matter gives them the necessary bond in ordinary firing.

From what we have established it is easy to feel that saving on the marl in the blend exposes one to breakage, distortion, *essuy*, &c. but if one is lavish with it, there will be loss of strength, and flaking, &c.

One does not find the same blend in in all the works. A considerable number of faience makers use equal parts of potter's clay and marl, or three parts of potters clay, two parts of coloured fireclay and five parts of marl. However, the differences found among various clays of the same type can produce even larger differences in the blends. All those that do not separate in very fine particles in water should be regarded as harmful. There many proper ways of assisting this division, [such as] freezing or stirring during a long immersion in water. Thus to avoid errors prejudicial to the blend, it is appropriate to make tests on the clays separately, to expose them to freezing, then humidity, to stir vigorously in water, to leave them there for a long time, and then to sieve them through a very fine sieve. We will see when we talk of enamels that the simplest and most certain rule is to use in the blend as much as possible of white earth without impairing the strength of the biscuit when fired for twenty-four hours in a well constructed furnace.

In the majority of faience works it is the custom to throw the three types of clay into a trough, leave them a certain time, blend them then tread them. I shall not stop to emphasize the inadequacy of this method to achieve complete sub-division and perfect blending of the clays. Describing what is done at Aprey is, I believe, sufficient to indicate what is to be preferred. M. de Vilhaut takes care to obtain his clays before winter comes, so that freezing will open them and make them finely divided. In the spring he makes his blend in a mud bath where they are broken down and mixed very well. On being taken from the mud bath the slurry is put into a riddle then carried by a long channel to a horsehair sieve from which it settles out in a very large basin which allows the water to be drawn off as the clay settles out. The riddle catches the largest particles, the sand grains of ordinary size settle out in the channel and the sieve catches what remains as fine scouring sand. When the clay has become somewhat firm it is transferred to another deeper covered basin from which it is taken to be trodden and then put into properly vaulted and paved cellars where it is stored to age and intermix for as long as the [rate of] consumption permits. As soon as the first basin is empty it is refilled again so that the clay may be exposed to the greatest rigours of the winter.

This blend being thus prepared provides a very good fireclay. if the ferruginous matter makes makes the biscuit too impervious, too compact for this type of faience, it is customary to choose a potter's clay that in which the ferruginous matter develops with greater difficulty and to add to the blend a sand of medium size.

I do not presume that there is a blend that will produce all the qualities that one would desire in faience which is better than that using equal parts of pure fireclay and pure marl, like that which is called white of Troyes. These two types of clay are not as rare as might be supposed; there are many ways of replacing the latter. this composition has only the disadvantage that it requires twice the heat of ordinary faience firing but one would be amply compensated for this outlay by the pleasure of seeing the white biscuit, having a faience light but very strong, capable of withstanding heat, beautifully white and admirably suited to receiving colours.

The production of the *white* or enamel is another essential part of faience manufacture. There is as much ignorance and prejudice on this subject as about the clays.

One sees here as little uniformity in the proportions as in the choice of materials. According to the majority of manufacturers the sand from Nevers and from Boue, not far from Besancon are the only ones to make a beautiful glossy white. They have only the property of being a little more fusible than the best sands because of the ferruginous matter that they contain. Some prefer as flux soda from Alicante, others that from Carthage, yet others *salicote*, and some kelp; this group prefers potash and that group *salin* or gall. There are even some who prefer sea salt. With materials so different in their natures, how can they produce one and the same effect? Experience shows the contrary. One hundred pounds of frit of lead oxide together with about one seventh of fine tin [oxide] for ordinary faience, or a quarter for fine faience, is sufficient to melt 100 pounds of good sand. Thus the batch for the enamel does not require any flux other than the lead oxide. In the present case sea salt and gall cannot be considered as fluxes, as I have shown in my *Memoir on the causes of bubbles in glass*. These salts produce a different and very useful effect in enamel, that is to remove the gross colouring principle. Without their aid the enamel would be a more or less marked yellow, more or less disagreeable.

The best soda from Alicante and the best quality potash are the worst for making faience because they contain too great a quantity of fixed alkali salt and too little of the *gall*. Enamels in which they have been used are yellow, lacking in gloss, and crack because they do not contain enough neutral salt to remove the gross colorant; they render the enamel too soft and they make it too glassy. I have observed this phenomenon more than once and the manufacturers have been more surprised than convinced that this was the inevitable result of the excessively good quality of the raw material. They would prefer to believe that they had been cheated by the vendors.

Soda from Carthagena, *salicote* and kelp, containing less fixed alkali and more *gall* have less serious effects. Although one can only add 25 to 30 pounds of these materials to each batch of 200 pounds, it is essential for the goodness and beauty of the white not to use them, not even a few pounds of sea salt as is the practice of some works. It only decreases the problem to some extent whilst increasing the expense without good reason.

There is another disadvantage of using sodas; they contain a great deal of the colouring principle which cannot be entirely destroyed in either the storage dish [French: *colombin*] or in the fritting furnace. Is it not sufficient to have destroyed the yellow colour that ordinary sand and lead oxide produce? It seems that those practising the arts have often been more occupied with multiplying the difficulties than removing them.

It seems to me to have been solidly established in my memoir on the *Perfection of glass-making*, &c. that gall, sea salt, Glauber's salt and sulphated tartar, when reduced to vapours carry away with them the gross colouring principle of the materials with which they have been combined. Faience works every day provide proofs no less clear.

Sulphated tartar or potash *gall* is less suitable for faience than the other two because it is only a little fixed by fire. Ordinarily sea or kitchen salt succeeds better and produces a better effect in equal doses than potash gall, or even that of soda, because the potash salt is already in tiny particles, already open to humidity, and consequently more disposed to combine with the other materials during melting than to rarefaction and evaporation. The *gall* [on the other hand] is in large and very compact pieces which it is very difficult to reduce to powder; it is deprived of moisture and carries a lot of colouring principle. This difference is even more evident when the manufacturers have not crushed it with care. I have very often seen lumps of gall as big as a pea [misprint: *poids* for *pois*] in bars of enamel, clear proof of imperfect combination and that the gall was not sufficiently finely divided to be evaporated by the heat and thus carry away the colouring agent. "The inconvenience is not as great as you think," you may say, "this gall will be pulverized with the enamel in the mill and it will have its effect on the ware when it is glazed." This reasoning has only the appearance of truth. The enamel is milled in water under a horizontal millstone; the water dissolves the gall which is then undoubtedly carried away when the water is decanted.

Soda-gall, suitably prepared, gives a greater effect, weight for weight, than common sea salt because the latter carries a certain amount of water and a greater proportion of foreign matter. One can confirm this by dissolution of the two salts. To gain the best advantage from the soda-gall it must be crushed, dissolved in water to precipitate its foreign matter, especially the gross colouring principle, using a little potter's clay dispersed in water; then the clear liquid decanted and evaporated to dryness then cooled. This salt, once more moist, should be mixed very carefully with sand to put in the clay storage tub, or with sand and grog for making the frit. Perhaps the manufacturers would find this procedure too long and onerous. In that case, although they would suffer disadvantages for doing so, they could be satisfied with crushing the gall as finely as possible then, a few days before using it, putting it in sufficient water for it to be penetrated and opened so that it is in the same state as sea salt when it is bought. This precaution is essential.

Water, as we have already said, disposes neutral salts to melt and to evaporate; their surface is increased by finer division and these salts like almost all others can only react when they are in contact. It is established fact that an old frit composed of equal parts of soda and sand becomes white on firing more quickly than a fresh mixture of the same materials in the same proportions. Why? because the moisture of the air or the place has had time to penetrate intimately into the material. The proof, which operates in the same way, is to sprinkle a fresh frit with clear water until all parts are slightly wetted before putting it into the furnace.

Glass gall is not rare in France: it would become so if all its possible uses were known. Small glass works where they use only red potash produce a lot of this material; those that know only kelp, even more. This salt at present sells for 6 to 8 pounds per hundred weight. If this salt became less common or dearer, a new resource could be found in the extraction of the salt from kelp, or even soda from Villeneuve or Pérols in Languedoc. Three pounds of tin oxide, or four of well-purified ordinary lime are a very good equivalent gall, because of the small quantity of fixed alkali that they contain. Perhaps the wisdom of the Government would allow those who have difficulty in making use of these resources to buy sea salt at an advantageous price. There would be simple and effective ways of preventing fraud.

One normally mixes 100 lb of sand with from 8 to 20 lb of gall, moistens it and puts it, in a dish for blending clay, under the furnace for firing the faience or in the ash pit. After removing it from the furnace, this mixture will, if properly made, and if the walls of the dish were not too thick, have become very white. It is easily understood that the sand becomes much whiter in a glass fritting furnace where it can be raked whilst under the action of the heat. It would then cost a little more in wood and manipulation but about one fifth of the salt could be saved. Eight to 20 lb of well crushed gall and 100 lb of grog are added to the bowl for the clay, as we have said above. Then this mixture, carefully blended, is put under the faience furnace in a new dish. If the gall has been prepared as we have indicated, 25 or 30 lb will suffice; for the rest a greater quantity cannot be harmful. the *white* can only be even better. those who do not use the special dish [to prepare their raw materials] are not to be imitated.

The proportion of 16 lb of fine tin [oxide] or 28 lb of common tin [oxide] for dishes to 100 lb of lead [oxide] seem to me very good for ordinary faience but 32 or 33 lb of tin to 100 of lead, as normally used for fine faience, seems to me too great, making flaking almost inevitable and producing a dull white. the enamel produced using the the latter seems to me too hard to *bite* sufficiently on the clay body made as described above and thus not to bond strongly to it.[a] One can, it is true, largely prevent spalling by requiring the workmen to sponge their ware only with *barbotine*, a mixture of very finely divided potter's clay and coloured fireclay, or not to sponge their work at all for fear that they make the clay *too lean* and leave on the surface only the calcareous part. This is how to be ruined by paying too much attention to the concerns of the workers. It appears to me that it would be wiser and more certain only to add 25 lb of fine tin to 100 lb of lead; the enamel would be very well bonded to the biscuit and of a beautiful white, tending slightly to blue which is the tint most sought after in faience. As a general rule it is less dangerous to decrease the tin in the enamel than the marl in the composition of the body. I believe that I have proved that.

Flaking reveals a singular phenomenon. Whenever the enamel flakes it is more or less bloated. What can be the cause of this swelling, this bloating? It seems to me that this can only be the due to a vapour which escapes from the clay in the last degree of firing. The enamel being too compact to allow it to escape and too firmly bonded to the body, gives way to it and in becoming distended to a certain extent, comes to a balance; but what is the nature of this vapour? The question is, in my opinion, very difficult to decide. Might it not be sulphuric acid normally found in the clay? I would believe that more willingly if I had never seen flaking with bloating on faience made with pure fireclay and pure lime.

It is not uncommon to see the red of the clay through the enamel, the layer of enamel being too thin. I assume that the enamel is not too soft and

---

a. See p. 408, No. 3 in the quarto Art of Glassmaking.

that it has not been pushed too far by the fire as these two causes could themselves produce the same effect as too thin a layer of white enamel.

Surface blemishes are the product of a thick smoke that has affected the biscuit or of carelessness of the workers who have touched the ware with greasy or soapy fingers. can one wait for these people to observe the necessary cleanliness? It is more prudent to prevent the results of their faults. This fortunately has been done by M. Vilhaut as far as redness and surface blemishes are concerned and the remedy is as sure as it is simple. It consists in grinding the enamel less [finely] than is customary, to use it at the size of ordinary sand grains. It is also common to see on faience *picassures*, black or grey pimples which are only tiny particles of lead formed when the enamel has not been sufficiently carefully purged of gross colouring matter.

Fine faience only differs from ordinary in the elegance of the shapes, the whiteness and brilliance of the enamel, the fineness and sparkle of the colours, and the beauty of the painting. Many things could be said about the colours but as they are basically the same as those for porcelain I shall defer to the lessons of my teachers, MM. Hellot and Macquer.

OBSERVATIONS ON CRUCIBLES FROM THE AUVERGNE

Communicated to the Royal Academy of Sciences, and that of Dijon, in 1771

IT is as important to have excellent crucibles as it is difficult to procure them. One cannot regard as such those that easily break either during drying or when exposed to the action of heat; those that have the slightest tendency to vitrify; those that soften and deform in a violent and long continued heat; that are too porous to hold fluid melts and which colour fine glasses or which impair the colour and the ductility of good metals. Which are the crucibles that have none of these defects? There is no artist who can make them and who has not had occasion to make complaints in this regard.

The crucibles that may be found in France at merchants or at potters are very far from having the desired qualities. No one can be ignorant that the best known and most esteemed are those from Hesse and Ipsen.

Do these two types merit their reputation and the expense that one must bear to obtain them? The first are so sensitive to heat that they cannot be raised to red heat without the very greatest care; when red hot, contact with the air or any material or a cold tool is sufficient to crack them; so for delicate operations it is rarely that one does not use two crucibles, one inside the other. Those from Ipsen are so disposed to melt that all molten fluxes attack them very easily; they also have the defect of colouring good quality glasses and of impairing to some extent the colour and ductility of gold or silver the first time that they are used[a].

---

a. See The elements of Docimasy of Cramer, French translation, vol. I, pp 343-7.

What means remain to the worker to obtain good crucibles? He can make them himself but this is tedious, difficult and uncertain. It is sufficient to read with attention the memoir of Pott[a] on this subject to be convinced. One would have to carry too far a love of the art not to be frightened off by all the precautions that this knowledgable chemist prescribes; how can one be sure of satisfying all the conditions that he believes necessary?

In my memoir on the improvement of glass making I believed that I had given the simplest and most reliable method of making good crucibles and I had reason to think so in relation to all the clays that I then knew; my method had one inconvenience common to all known methods, that of requiring a blend of clays. Although my pots required only the blending of two materials, fresh clay and grog, which differed only in one of them being calcined, their quality was always uncertain up to a certain point.

It is only since I have known the white clays of the Auvergne that I have been convinced that the inconvenience I have just mentioned is not inevitable. These clays do not require any addition to produce excellent crucibles: thus what I have to place before the eyes of the academy will appear less prodigies of art than due to the riches of nature.

White clay is common in the Auvergne; I have found it in the beautiful forest of Montel de Gelat, two leagues from Pont-au-Mur, on the road from Clermont to Limoges, in many places near the way from Langeac to Saint Flour, around Malzieux, at Javogue, two leagues from Brioude, at Bordet and one league from Saint-Germain-Lambron, on the road from Clermont to Brioude; between Usson and Sauxillanges, a league and a half from Issoire, at Marsac not far from Riom, at Bordpré near Lezon, on the road from Clermont to Thiers; there are no open mines for the last four. I do not doubt that there are good white clays in many other parts of the province; chance or new investigations will make them known. This white clay is usually found beneath a layer of sand of variable thickness and colour.

---

a. See the Chemical Dissertations of Pott, 4th vol.,p 167.

This clay is not found in the ground like the white clay of Boëlu near Chimay or like the grey clay of Belière in Normandy, in [large] masses or isolated bodies but rather in reefs of a certain extent and of a thickness that still remains to be determined.

It is found at different depths, from three to twelve feet, occasionally up to sixty. These white clays, excepting that of Bordet, are mixed with sand and mica; the sand that they contain is very white with sharp corners and the grains are of widely differing sizes, thus there is reason to believe that the sand comes from crushed quartz. There is none of these clays which, when deprived of all its sand and mica, does not appear to have lost some of its *gluten* and, in this state it does not spit or crackle when thrown on hot coals; also it decreases very little in volume when exposed to the most violent heat. Far from manifesting the slightest tendency to vitrify it does not even show a surface film and acquires the greatest degree of whiteness.

Because of its singularity, the white clay of Bordet merits a special description. It is found at a small depth; the reef appears to be extensive and very thick; it is extracted in lumps almost as hard as ordinary quarried limestone; it is of the greatest fineness, very pure, without any admixture of either sand or mica. It appears to have lost all its *gluten*; when reduced to a very fine powder and kneaded with water it has no plasticity; it is impossible to work on a wheel; it is unchanged by heat.

When I arrived in the Auvergne crucibles for the apothecaries, founders, and goldsmiths of the province and of the town of Puy in Velay were made at Javogue, Sauxillanges, Marsac and Bordpré. Although the pot makers of these four neighbourhoods did not take appropriate care with the clay, purifying it and removing the coarsest sand with little care, the crucibles that they made without any other preparation or any type of blending, can be regarded as the best that one could find.

I subjected a crucible from Sauxillanges to a very severe test similar to that which I had used for a sample. Full of glass, hanging over the rim of a large pot and supported at the top by resting against the wall of the furnace, it was subjected to the heat of the glass making furnace for sixty three days without appearing to have suffered at all by distortion of its shape or loss of substance. It would have lasted much longer if the founder had not clumsily made it fall into the furnace whilst *charging batch* into the large pot.

Last year I had some glass making pots made in wooden moulds from white Sauxillanges clay extracted and purified with care but without any admixture. They were oval in shape, the major diameter being twenty inches and the minor twelve. This prepared clay is so little unbending, so little disposed to cracking, doubtless because of its lack of *gluten,* that I could dry them in the sun and could even do so before they had been removed from the moulds. Having been tamped gently but frequently they showed no fissuring. Pots made from this clay are very fragile and lacking in strength when dry, requiring to be handled very carefully. It is only after exposure for some time to a very violent heat that they acquire the greatest strength, to the point of striking sparks in a tinder box; in this state their fractured surfaces are of an astonishing whiteness.

At the beginning of August last year I fitted out a melting furnace with eight of these pots; there are still some which appear to have suffered scarcely any damage from the action of heat or of the fluxes. I would observe that the oval shape is obviously unfavourable to the life of the pots but is advantageous to the production of furnaces making miscellaneous ware.

I do not think that it would be prudent to make pots of a considerable capacity from this clay without any addition; they would be too heavy to be handled with the necessary care. Glass houses making bottles, sheet or plate glass, &c., would, I am sure, like to be given a means of using our white clay for their large pots, to obtain almost the same advantage. It is only necessary to mix with this clay, as precisely as possible, one fifth of the purest, finest, richest, grey or brown clay; these clays are common in France; this procedure has been very successful for me. The fifth of fresh clay is sufficient to give large pots the strength needed for them to be removed [from the moulds], moved around, fired, and transferred to the melting furnace.

On Glass Making

At the beginning of last year I had a melting furnace built with white Sauxillanges clay, without being purified, such as has been obtained from Puits; it was dried rapidly and showed very little shrinkage. Although this furnace has been in service for more than eight months, it does not show signs of wear. To judge by its present state and the amount of working done in it, one can believe that, with some minor repairs, it will last for ten to twelve years[a]. This type of furnace made in the ordinary way used by glassworks would last only fifteen to eighteen months.

I may be asked why the white clays of the Auvergne do not require any addition or intermediary, that no fired material needs to be mixed with them to make pots and excellent melting furnaces when all other known clays do require such additions. The reply is simple & I do not need to explain it to the Academy. The other known clays are too pure and too rich for the water, and perhaps the acid, that they contain to be released only with great difficulty, so that it is consequently turned to vapour between the outer layers by the least degree of heat. This inevitably causes a continuous dissolution of pots made from them. However, the white clays of the Auvergne always contain some mica and a little fine sand and they have much less *gluten* than the other clays; thus they have much less need to be blended and they can be used in their natural state.

I believe that Auvergne crucibles could, without doubt, be improved; I am convinced they become better as more care is taken over the extraction and cleanliness of the clay; it is rare not to find the clay to contain some small roots and foreign earths. As normally purified the roots are usually removed but the finely divided coloured earths remain and dispose the crucibles to vitrification.

The methods used by potmakers of the Auvergne to purify these clays are as uncertain as they are costly. Some, having suspended the clay in water, pass it through a cylindrical sieve of a coarse mesh which is turned on two supports by means of a crank; this operation is tedious and allows

---

a. [Later note]: This furnace was worked for 27 months without repairs.

a great deal of clay to be lost without preventing all the foreign bodies falling into the vessel which collects the clay from the sieve. Other are satisfied with mixing it with water in a tub or barrel and transferring it from one to another when they judge that the coarsest sand has settled out, without giving any attention to removing the light bodies that float on the surface: (it seems useless to me to point out that many bodies remaining suspended in flowing or agitated water do not remain so in still, clear water). A day or two after the operations just described, the workmen decant the clear water and put the purified clay to dry on large wooden tables, under a partly covered shelter where it may collect dust, feathers, and foreign bodies; from there they take it to the workshop in poorly closed chests.

On first seeing this method I was astonished by its defects and regretted seeing so precious a material treated with so little care. I would have been pleased to put my insight and experience at the service of the pot makers but I did not find them disposed to profit thereby. One from Javogue told me that he had practised his art for much too long a time to need any advice. The only pot maker in Sauxillanges was more amenable and his pots were, in fact, the best because he extracted and purified his clay with greater care.

I do not believe that I need to repeat here all the details that I have given, in my *Memoir on glass making*, on the special care and extreme cleanliness needed with pot clay, in extraction, transport, purification, drying, kneading, and storage. The method of large scale purification that I have given there is simple, reliable and cheap: it consists of mixing the clay with a large amount of water in a large tank then running it off through a hole in one side of the tank onto a sieve of horsehair or very fine bow hair and into other tanks, covering the tanks carefully until the clay is completely precipitated, then drawing off the clear water through openings at different heights in the sides of the tanks.

Potters who make crucibles with the white clay also make other vessels from ordinary potter's clay in the same workshop. It is very difficult, not to say impossible, to avoid mixing the two clays. To make our pots as perfect as they could be, it seems to me necessary to have a workshop especially for them and [that it be] carefully maintained.

It would be almost as useful to be able to guide the hands of our potters, or at least persuade them of the importance of throwing and maintaining their pots true; of making them a uniform thickness around the whole circumference, of making the interior shape as true an inverted cone as possible, of trimming them, of uniting their parts with the greatest care and of matching them so that they can be packed, stood one inside the other, without knocking together. It would be desirable to provide a slight flare on small crucibles or, perhaps even better, that supports to the depth of about quarter of an inch should be provided where the bottoms of the small crucibles are nested together; by this means one can avoid the too frequently seen inconvenience of crucibles toppling over in the furnace. Users need crucibles of different sizes from an inch and a half up to ten inches in height. There is always room to hope that pot makers desiring to make a large sale in their own interests, will pay attention to our advice and be disposed to profit from it.

The white clay of the Auvergne does not only produce excellent crucibles. It can be used for muffles, cupels or ladles for slag or vitrification; supports, crucibles for cementation and their lids, retorts of good quality, all clay vessels needed for the physico-pyrotechnic arts; it is only necessary to provide our potters with good models.

All the makers of crucibles are, more or less along the banks of the Allier; a very fortunate situation for sending their products to nearly all parts of France, above all to Paris, and also abroad.

We could thus arrange to stop a considerable importation, to have from here all our own clay vessels needed by chemists, assayers, enamellers, founders and goldsmiths, of a better quality and at a much lower price than those we now obtain from Germany, Holland, &c. I do not doubt that in a short while the province of the Auvergne will supply nearly all the glass works of the kingdom from its white clay, that its manufacture of crucibles will become as flourishing as those of Ipsen, Hesse, Valdenburg, &c., or that the greatest advantage will be obtained.

At the present time this manufacture needs the protection of the Government, above all against the greed of some proprietors in areas where the white clay is found.

M. de Monthyon, whom this province has the good fortune to have as administrator, is too far sighted and has the best interests of the province, as well as the progress of the arts and sciences, too much at heart to refuse it; he has already made arrangements very favourable to them. Why should useful establishments have to wait for whatever can anticipate and prepare for their needs?

We conclude this memoir with two observations, one related to the nature of the white clay and the other concerning pots.

I have observed that our white clay is as favourable for plants as other pure clays are known to be the opposite; various seeds that I have sown in it have grown with vigour. However, I believe that Kulbel would have found much difficulty in extracting from this clay the volatile salt that he regarded as the principal cause of the fertility of the soil.[a]

The surest means by which we can make our pots for *melting lead glass have a long life* is to *glaze* them with ordinary white glass, the batch for which contains no lead oxide. If one is not already melting glass of this kind, it is sufficient to coat the inside surface of the pot with such glass reduced to powder and then to allow it to melt before putting into the pots the glass or batch containing oxides of lead. This glass, with a more viscous flux, prevents the more vigorous one attacking and rapidly destroying the pots through their pores.

---

a. See Kulbel diss. de sel volat. causa fertil. terrarum.

MEMOIR ON THE FALSE EMERALD OF THE AUVERGNE

Read to the Royal Academy of Sciences and communicated to that of Dijon at the beginning of 1771

THERE is probably no province in France that offers such natural riches as the the Auvergne; in almost all parts, either the lowlands or the heights, one finds rare minerals fit to whet the curiosity of true physicists. One of those that has interested me most is the false emerald of Loubeyrac, land belonging to the Marquise de la Fayette. False amethyst is also found in the neighbourhood of the Abbey of Pebrac but we are only concerned in this memoir with the false emerald*.

It is known that this stone is a fusible or vitreous spar. When struck it fractures into irregular pieces like glass; it can be scratched with a knife; it does not effervesce with acids; it makes sparks in a tinder box, in a very obvious manner, at least in the dark, although Wallerius and Pott have stated that fusible spar does not strike sparks with steel[a]. Our false emerald appears to be a true crystal, its crystals being of cubic form, large, well formed, very transparent and clear green; it has a greater specific gravity than quartz [but] I will not agree with Pott[b] that on this account it must contain mercurial matter, about which I admit my ignorance. With respect to its metallic constituents, I have made without success many attempts to identify them. One of the characteristics said to be typical of fusible

---

* The mineral is later identified as fluorspar or fluorite, CaF_2.

a. Mineralogy of Wallerius (French translation) vol.1, p.123. *Chemical Examination* of Pott (French translation) p.166.

b. See above, p.164, concerning mercurial matter.

spar is to crackle and decrepitate as soon as it is exposed to heat[c] the false emerald of which I speak possesses this property markedly; the matrix of the crystals bleaches very easily on heating and does so rather more rapidly than the crystals. Pott believes that this non-crystalline matrix is of quartz but it is as soft as the crystals which also strike sparks with steel.

These crystals become luminescent [D'Antic says *phosphorescent*, the distinction was not then recognized] if exposed to a low heat; a very agreeable spectacle may be observed by gradually exposing the false emerald of Loubeyrac to the heat of a kitchen fire; as soon as the heat has penetrated it, but before it becomes red hot, one can see with pleasure the luminescence develop in the best crystallized and most transparent parts, revealing itself by a pale blue flame which will impress even the least sensitive people.

Decrepitation

If one moves the false emerald nearer to the fire it will suddenly decrepitate and an enormous number of pieces which look like tiny particles of burning sulphur fly off. One can also produce a very interesting type of firework by throwing some crystals of our false emerald on hot coals; what is particularly unusual is that these phenomena are almost as visible by day as by night. The green fusible spars from Memmendots, Giromagny in the Vosges, and Bourbon-L'Archambault, which I found in the rich natural history collection of M. Varenne of Beast, produced the same effects.

This phosphorescent effect ceases on heating; it also stops with the decrepitation but the decrepitation only ceases when the green colour disappears and the crystals have become almost perfectly white. (*Marginal note*: Both of these properties cease as the false emerald becomes hot but the decrepitation only when the green colour has disappeared.) On observing these phenomena carefully it did not seem to me difficult to explain the phosphorescence of the false emerald and consequently of all

---

c. Wallerius, p.124.

coloured fusible spars; it can clearly be seen that these crystals take on this property because of the every dilute and volatile colouring agent which gives them their tint and which is made mobile by the heat, hence the false emerald loses its phosphorescence with its green colour. However, I believe that I have proved in my memoir on glass making and in a more extended form in my memoir on electricity that the colouring agent is nothing other than phlogiston, the inflammable principle.

The false emerald of Loubeyrac is the phosphor of Balduinus and the lithophosphor of Woodward[d].

Everything that we have said up to now about these crystals seems to me merely interesting, it is not on that account that it intrigued me so much, it was on account of the advantages that man can obtain from it. I have attempted to make it useful to the arts; the Academy can judge how far I have succeeded.

Ideas on the uses of fusible spar

Very few authors have spoken in any detail about fusible spar; the mineral is hardly known except in natural history collections of people who work with ores and to metallurgists. Experienced workers in ores know that it is an indicator of good ores, that it facilitates melting, above all of those that have lime-rich materials as gangue, without doubt because it has more affinity with lime minerals and earths than any others[e]. Pott was, I believe, the first chemist to pay attention to fusible spar in his *Lithogeognosy* where he examined its reactions in relation to heat, salts, glasses, and many minerals or oxides; every one should consult the work of Pott[f] to appreciate the value of his experiments.

Accepting the word of Pott, I used to believe that fusible spar would not melt [even] in a violent fire but chance showed me that he was mistaken. Intending to calcine some strongly, I put a fragment of false emerald into

---

d. see his vol.2.

e. see Pott, *Chemical Examination of Stones*, p. 202.

f. see Pott, *Chemical Examination of Stones*, p.171.

a crucible in a small glass-making furnace. A quarter of an hour later, although I had not even crushed it, I found it completely melted and almost as fluid as oil; the glass that it had produced was well refined, solid, much denser than the poorest bottle glass, coloured green with a slight yellowish tint, as transparent as good quality ordinary window glass; a little time later I found no glass left in the crucible, it had eaten away and vitrified two thirds of the base and had penetrated right through. I repeated the experiment in a crucible made of very pure clay properly compounded but it was equally attacked although its exterior was not even glazed because the clay was so refractory.

Non-crystalline material less fusible than the crystals

I have separately tested the crystals and the matrix which contains crystals of false emerald. I did not find it possible to see any difference between the glasses; I observed only that the non-crystalline material was a little less fusible and gave a less strongly coloured glass than the crystals.

The glass more corrosive than lead glass

In all the experiments that I have made with this mineral, without any addition, I have consistently found that the glass attacks the crucible more energetically than lead glass. The experiments of M. Pott[g] however demonstrate that clay is not the earth with which the spar has the greatest affinity.

It colours very well glasses in which it is used

The slightly yellowish green tint seen in the false emerald indicates a grosser and more fixed colouring agent than that with which its crystals are tinted. This agent is so fixed and so abundant that, when used in glass batches, it colours the glasses more than the same quantity of slaked lime. The use that I have made of this observation will be seen later.

I have added different proportions of false emerald to batches for white

---

g. See above [ref. f.]

glass. These experiments have given me opportunity to make several interesting observations which I shall now recount.

Effects of mixing it into glass batch

In whatever proportions, up to one tenth, that one adds these crystals to glass batch, the glass is semi-transparent and milky (like a milk or good chalk [opal] glass). If a little manganese is added to the batch one gets a very agreeable opal. However, the action the heat used for melting makes the milkiness disappear and the glass becomes completely transparent but always with a greater or lesser degree of green colour in proportion to the amount of false emerald added. When one reworks this glass the flame to which it must be exposed makes the milkiness or opalescence reappear, just as happens with a glass containing too much lime. Glass made only from the false emerald shows the same changes.

A means of making opals

We thus have a new means of making, at little expense, milk glass and artificial opals which are harder and less fragile than those made with chalk, slaked lime, or calcined bones or stag's horn.[h]

Frit mixed with the false emerald impedes corrosion of the pot

False emerald mixed with *frit* in effect loses its corrosive nature; the pots then seem to be corroded no more than by glasses made from normal batches.

Porcelain made with glass containing false emerald

This glass, which becomes milky in a flame makes an excellent glass porcelain when sintered together with lime, plaster or sand. It has all the good qualities of that [invented by] Réaumur without possessing its defects: it has a better gloss, the whiteness is more pleasing and its surface does not show so great a number of dimples or cavities. The celebrated

---

h. See Neri's *Art of Glass* (French translation) p.103.

Réaumur believed that his porcelain would improve appreciably in quality as the quality of the glass used in sintering improved; what we have just seen about false emerald glass is not the only evidence that we have to the contrary.

Effects that could be produced in glasses if melting did not produce opalescence

It is annoying that glass in which these crystals have been used, in certain proportions, loses much of its transparency in a flame and becomes milky. In batches for fine quality glass it would save the always expensive fixed alkali by facilitating purification and making it less fragile; in ordinary glass batches it could take the place of the majority of the soda and ashes that are used, contributing to their strength and their refining; in batches for bottle glass it would shorten the time needed for melting, make the glass less easily broken by cooling and easier to work; three objects of great interest in a bottle glass house.

Advantages that may be attained in glass houses

Not, however, believing that our false emerald should remain useless for the three types of glass mentioned it could be used to convert bottle glass in glass porcelain for chemical vessels or cooking vessels, by substituting it for one quarter of the leached ashes; this would, at once, provide a well refined glass, sweet to work, less fragile, and very easy to convert into glass porcelain, if the soda and the ash were strongly fritted before using them in batches for ordinary glass.

It can replace soda and ashes

The false emerald can replace one twelfth of these two fluxes and contribute to removing bubbles from the glass and improve its quality. With regard to fine white glass, it has considerable advantages. Up to one twentieth may be used in batches for this type of glass without fear of diminishing its transparency or impairing its colour.

It can replace lime

This [one twentieth] is the proportion of slaked lime usually used to make the glass more fluid, less viscous, to facilitate its purification, and to prevent defects due to gall.

False emerald can produce exactly the same effects but with the difference that slaked lime requires an increase, about equal to its own weight, of fixed alkali and disposes even sand or flint to melt. If by negligence of the *teaser* or bad quality wood, or a deficiency in the batch, the glass becomes fatty, frothy or cloudy, one can dispel the defect by throwing a certain quantity of powdered false emerald into the pot, and do so more quickly and more successfully than by using arsenic, antimony, or powdered coal, green wood, &c., and without fear of producing streaks or making the glass badly refined. It only produces this desirable effect through its strong colouring agent. Gall, sulphated tartar, and Glauber's salt, which are the sole cause of smears, having greater affinity for the colouring agent, the phlogiston, than with the glass, or what comes to the same, the colouring agent having greater affinity with the gall than with the glass[a], the colouring agent leaves the latter to combine with the former. As a result of this combination the gall becomes volatile and the action of the heat purges the glass.

It can be useful in porcelain, earthenware, and bricks

I am persuaded that these are not the only advantages that can be obtained by using the false emerald that I have just described; there is every reason to believe that it can be used in porcelain, either for the body or the glaze, by makers of stoneware, potters, and even by tilemakers to make bricks, glazed tiles, &c., &c. All these objects enter into a plan for experiments on this mineral that I have made for myself.

When I had the honour to read my memoir on the false emerald of the Auvergne to the Royal Academy of Sciences I did not know, and could

---

a. I believe that I have proved that in my memoir on *Smears*.

not have known, of the beautiful discoveries made by Marggraf, Scheele and Sage concerning fusible spar.

These scientists proved that fusible spar is composed of lime and an acid; that this acid attacks and decomposes glass and makes fixed alkalis caustic, deliquescent and gelatinous, that is to say that it removed their *aerial* acid [carbon dioxide].

This [hydrofluoric] acid precipitates out soluble silicates by decomposing the glass and probably regenerating the quartz-like constituent (the film formed in the receiver of the acid during distillation is certainly quartz-like); forming on long digestion with selenite earth true rock crystals, it seems to me difficult not to believe that it is a constituent part of earths and rocks of flint-like type[a]. Achard has succeeded in forming rock crystals by an ingenious method, which has been successfully repeated by Magellan, and involves combining aerial acid with pure clay or the oxide of alum. M. Schlosser, Doctor of Medicine and member of the Royal Society in London, has evidently proved in his excellent treatise on the essential salt of urine, as M. Margraff had demonstrated for the acid of fluorspar, relative to fixed alkali, that the acid of urine removes their aerial acid from ordinary salt of ammonia and from subliming alkali and gives this acid its liquid nature and fixity. Can one not conclude from this beautiful experiment that the aerial acid only differs from that of urine by its elasticity, by a simple modification? But no-one doubts that the acid of urine is phosphoric acid. It thus seems to me incontestable that the acids of fusible spar or urine, aerial acid and phosphoric acid are one and the same acid, as Sage has claimed.

The knowledgeable chemists who have made such curious and interesting experiments on fusible spar do not appear to me to have recognized that the phosphoric acid is present in vitreous spar in two states: one combined with phlogiston in the true phosphoric state which can be evolved by only a slight degree of heat and another that neutralizes the calcium in a state of purity and fixity, which cannot be evolved by even

---

a. See Bergman Col. 26, Terr. silic. de attract. elect.

the most violent heat, and which has the advantage of a double affinity for sulphuric acid.

The phosphorescence of vitreous spar thrown onto hot coals is, in my opinion, obvious proof that part of the acid, at least a large part, exists in this mineral in the state of true phosphorus.

It is no less certain that having undergone this test, it can still, by means of equal parts of oil of vitriol, produce a large quantity of pure or almost pure acid, because it is combined with the inflammable matter in the sulphuric acid, inflammable matter which suffices to give it a certain degree of volatility but is insufficient to make it luminous.

OBSERVATIONS ON THE MANUFACTURE OF AND COMMERCE IN POTASH

Compiled in 1775 at the request of the Society of Arts in London

THERE are two types of potash, red and white. Red potash, also known as *salin** or *glass ash*, is the yellowish and greasy fixed alkali obtained by leaching and evaporation from the ashes of all woods and plants except maritime ones. White potash is nothing other than red potash calcined until white or until it has taken on a bluish tint.

Red and white potash which contain little neutral salt, or sulphated tartar [K_2SO_4], are called *lean* potash and those that contain large proportions are called *rich* or *fat*. They are given these descriptions because glass made with the latter is cloudy, *fatty*, whilst that made with the former is usually yellow coloured. To leave out nothing of importance about the manufacture of potash it is appropriate to discuss: 1) Which woods and plants, when reduced to ashes, give the greatest amounts of fixed alkali or the greatest purity? 2) What is the most suitable time to burn them? 3) How should they be burnt? 4) What precautions should be taken with the ashes before leaching? 5) How should the leaching be done? 6) How should the evaporation or production of the red potash be conducted? 7) What care is needed in the conversion of red potash into white?

1. The different types of wood that give the most ash, are those, other things being equal, that are the densest, hardest, or heaviest. Thus box, oak, hornbeam, beech, &c. give much more ash than pine, fir, willow, birch, &c.

Trees that are stunted, ringed, pollarded, full of excrescences, or internally rotten give much more ash than those that are sound and well grown. It seems that nature in her wisdom has chosen to compensate

---

* *Salin*: used by glass makers for alkali in general (Littré).

mankind for the losses caused by accidents or maladies of trees.

Worm-eaten wood or natural bark taken before the tree is completely dried out, or dead but still standing, are almost entirely converted into fixed alkali salt. It is almost the same with excrescences or gnarled lumps, especially on resinous trees. These two observations that I have made over the course of more than ten years would alone prove that those who have, up to now, laid down the theory of nature and the formation of artificial fixed alkali are greatly in error.

I have observed that ashes of roots produce more fixed alkali than those from the trunk which in turn give more than the branches and the branches more than the leaves.

Of all the plants the fleshy leaved and leguminous ones are those that give the most fixed alkali.

Soft and white woods such as pine, fir, willow, hornbeam, beech, &c., &c., give the purest fixed alkali containing the least rich matter or colouring agent which is thus the easiest to extract from the ashes and recover by evaporation. One constantly finds the opposite properties in the ashes of oak, horse chestnut, sweet chestnut, &c. For this reason it is very desirable not to leach these ashes by themselves but to mix them with others which contain less colouring agent and to prepare the ashes before leaching as will be described in the appropriate place.

2. The oldest trees are the most appropriate for conversion into ashes. However, the same is not true of plants. To wait until they are fully mature will cause a distinct loss; it is necessary to cut and burn them before any parts of them begin to turn yellow. As soon as the green begins to fade a part of the vegetable acid, so necessary for alkali production, is dissipated and, when reduced to ashes the plants produce much less fixed alkali. The moment when the seed begins to form is the most appropriate to gather or cut the plants and convert them into ash.

It would be more useful to convert trees into ashes during autumn and winter than during spring or summer because they contain less water during these two seasons and the vegetable acid is less volatile but this is difficult because the water from snow or rains during the winter washes away the fixed alkali from the ashes either immediately after combustion or during transport.

3. To conduct the incineration properly it is important to choose a suitable place, dry and protected from the wind; to make as large a pile of wood and plants as possible, above all in height; to allow the plants and trees to wither somewhat; to make them burn slowly and without much flame, avoid heavy rains, and to make use of a period under cover, perhaps with a little rain; to collect together carefully and gently, with a wooden shovel, into the centre of the fire the ashes and embers from around the circumference; to allow all the embers to be reduced to ashes and to assist this process by gently stirring the fire from time to time; to cover the heap of ashes until one can remove them for fear of gusts of wind, rain or frost when the combustion has been carried out on a hillside. Wood and plants give a greater quantity of ash if they are burnt very slowly but they give a much purer alkali containing much less colouring agent and they evolve it much more abundantly when they have been exposed to a considerable and long continued heat.

Ashes of bracken made without care give, like kitchen ashes made by burning only green wood, only about one tenth their weight of fixed alkali; those made taking the precautions we have just detailed produce up to one third of alkali.

There would be a gain in both purity and quantity of fixed alkali if the ashes were incinerated again, and covering them with a pile of wood or plants which will be burnt subsequently.

To convert the bark of decayed trees into alkali it is only required, at any season, to choose a dry spell and set fire to the worm-eaten wood at the base of the tree; it will take hold promptly on all the decayed wood and these ashes will be also entirely fixed alkali salt which will fall on to the roots from which the earth should have been carefully moved aside. This method is capable of yielding a very large production in Canada and in North America where there are immense numbers of large decayed trees and they can be carried much further than those to be completely converted to ashes.

4. Even hot water dissolves the fixed alkali in new ashes [only] with difficulty, doubtless because they contain too much fatty matter or colouring agent and, for the same reason, the salt extracted is impure and much more difficult to convert into good white potash by calcination. Calcining furnaces are not the only ones which allow this phenomenon to be observed; it manifests itself in a more striking manner in melting furnaces. Whenever one uses this fixed alkali, without it being calcined, in batches for miscellaneous colourless ware, as is the custom in Bohemia, Alsace, and the whole of France, its fatty matter causes frothing and blisters in the glass early in melting; it is this potash that glass-makers call *lean*.

The people most experienced in these matters claim that ashes give more alkali as they are aged. It is clear that when prepared by the method we have described they undergo a decomposition and recombination, the excess of fatty matter disappears, the extraction of the salt becomes easier, and sulphated tartar [K_2SO_4] is formed.

This important [stage of] preparation consists in passing the ashes through a horsehair sieve to remove any [coarse particles of] carbonaceous matter which remain; to spread the sieved part in layers six inches thick on a floor properly paved with stone slabs, or lacking stone, floored with large and carefully jointed wooden planks; to sprinkle each layer with water and tamp it down; to raise the layers to height of four or five feet; to turn them over, sprinkling with water carefully from time to time and to keep them in this condition for six months. This is called *germinating* the ashes, or *allowing to germinate*.

The remarks just above and in the second paragraph of this memoir suggest that this is *rich* or *fat* potash, not *lean*.

A large number of experiments have convinced me that a considerably larger quantity of true sulphated tartar is formed when the ashes have been allowed to *germinate* for a long time even though the water used to dampen them does not contain an atom of selenite [$CaSO_4.2H_2O$]; see the notes added to the memoir which won the prize of the Royal Academy of Sciences*.

---

* Memoir on the perfection of glass making

5. Nothing else is as convenient as coppers of medium size or barrels for leaching the ashes. They should be raised fifteen inches above the floor of the workshop to allow the liquid to be drawn off into a bucket without difficulty. A second base with open joints is placed about two inches above the first and then covered with about three inches of straw. The vat is filled with ashes to within about an inch of the top, sieving them through a large horsehair sieve so that the water will dampen all parts equally. When the vat is full the ashes are tamped down so that the water will not drain through too rapidly. Only sufficient boiling water to moisten the ashes is added; an equal quantity is added two hours later and this process continued only until as much water has been added as is needed to dissolve all salts in the ashes. The whole is left in this state for three times twenty-four hours and the wooden stopper then withdrawn and replaced by one of straw, the lye is then run off into a bucket. If this operation is done with care it will be sufficiently concentrated to be worth evaporating.

It is not possible to specify the quantity of water to be added to the vat; it depends on the purity and quality of the ashes. Experience will teach what it is important to know in this respect. Some salters judge that the lye is very good when an egg will float in it. The majority test it by taste; others by the weight of insoluble ashes left on the filter mat but one must be sure that they are not contaminated with sand.

When all the lye has drained out the wooden stopper is replaced and the vat filled, as the first time, with hot water; [however] this lye would not be worth the cost of evaporating it.

It is used, instead of pure water, to wet new ashes put into barrels and produces an excellent effect. It contains a certain amount of fixed alkali and is better than pure water for dissolving the salt out of the ashes.

6. The apparatus needed to extract and precipitate the fixed alkali salt consists of a furnace* constructed of suitable stone# or bricks above which

---

* An improved type of furnace for this purpose is shown in fig 4, and described in the accompanying notes, see p. 73.

\# The normal translation *flint* for *pierre à feu* seems inappropriate here.

are placed, about twenty inches above the floor, along its length, at least three cast iron boilers each 30 to 36 inches diameter and 9 or 10 inches deep which are sealed in so that the flames can only pass out through the chimney which should be placed at the opposite end to the grate or *stoke hole*; a small iron shovel with a wooden handle to stir the salt; a rake to remove it and various wooden boxes to receive it; a chisel and a mallet to remove the salt when one has imprudently allowed it to stick to the bottom of the boiler. The first boiler, or that nearest the stoke hole, is used to reduce the lye and form the salt; the second to hold lye to replenish the first and the third holds water alone to be used for addition to the leaching vats: the process occurs more quickly when the fire is bright and well fuelled.

When the lye begins to thicken it is necessary to stir it almost continuously with the iron shovel. This movement assists evaporation, prevents boiling over and also the sticking of the salt to the bottom of the boiler. One may also, and this most advantageous and least arduous, keep the rake on the bottom of the boiler and put the salt into a box as soon as it is precipitated in sufficient quantity. When there is enough salt to make one batch the boiler is again filled three quarters full whilst the salt which has been put into the box is stirred continuously until it is completely dry. In this state, after it has cooled, it is put into a barrel and sealed to exclude all moisture.

It is essential not to put cold water into hot dry boilers, they may crack. If by negligence salt is allowed to stick to the base of the boiler it should be removed with the chisel or, if there is an appreciable amount, made to dissolve in fresh lye.

7. A furnace eight feet long, seven feet wide and three and a half feet high to the keystone, with an elliptically shaped crown sprung one foot above the *hearth,* which has a grate along one of the sides parallel to the mouth, 15 to 16 inches wide, constructed from hard bricks, the floor of which is carefully made of the same bricks set on edge, is the best kind for converting red potash into white. It is of the greatest importance to sweep both the floor and the crown of the furnace carefully with a strong

broom and to make it red hot before throwing in the salt. At one time seven to eight hundred pounds of red potash can be calcined.

The calcining furnace of which we have given the plan, section and elevation see Fig. 5 in our memoir on glass making, is equally suitable.

In front of the furnace it is useful to have a properly paved area made of stone[#] or hard bricks enclosed on three sides by large stones 18 to 24 inches high and properly jointed in which to deposit the potash and allow it to cool when it is withdrawn from the furnace with an iron shovel shaped like a scoop; when it is cold it is put into barrels or properly jointed boxes, whichever are the best value and most convenient, taking care to tamp it down well with a piece of timber. If it is intended to go overseas it is very useful to tar the outside of the barrels or cases to exclude all moisture.

The conversion of red potash into white is a very simple operation. Three main points need attention; firstly to begin calcination with a low fire and not to stir the salt until the moment of *aqueous fusion* has passed. When red potash contains a great deal of fatty matter, of colouring matter, it releases its water of crystallization with difficulty. If the fire is increased too vigorously this water is evaporated suddenly, increasing the relative quantity or volume, it gives the salt too great a fluidity which is troublesome, allowing it to flow out of the front of furnace, or easily forming a *cake*, and makes it very difficult to calcine. This difficulty is hardly to be feared when the salt has been extracted from prepared ashes, as has been indicated above; by this method the fatty matter has been decreased and a good part of it has been dissipated because it has made the salt easier to dissolve and, as an inevitable consequence, easier to dehydrate.

The second point is that the flame must be clear, rapid, and well maintained, [otherwise] one can see that the smoke would return to the salt what one is trying to remove from it.

Finally, the third point is that the entire surface of each particle must be exposed to the action of the flame by stirring it continually, turning it over with an iron rake with large prongs and with a flat iron shovel,

---

[#] Again *pierre à feu* but not likely to mean *flint*.

spreading it out and then making it into a heap again so that the material on the bottom is frequently carried to the top and vice versa, and by crushing the very small lumps with the handle of the shovel or rake, &c. &c. When the potash, slightly cooled on the shovel, is very white or appears slightly bluish, it is removed from the furnace when still red hot and new red potash is charged to avoid wasting the heat of furnace.

We cannot recommend cleanliness too strongly. It is also essential that no particles of clay or stone from the furnace, nor any iron scale from the shovel or the rake, get mixed with the potash.

Commerce in potash

In Alsace and Lorraine there is an abundance of merchants who buy red potash from small manufacturers and then convert it into white potash. This practice is very advantageous to the trade.

The experiments and observations on American potash of Mr. Lewis, member of the Royal Society of London, and the observations by R. Dossie on the excellence and the adulteration of American potash have been dignified [with publication] by those authors; they were made by the hands of masters, but it is to be feared that they will be more useful to chemists than to manufacturers or merchants of potash; they are too abstruse for ordinary men to be much instructed by them. Without departing from the principles of Lewis and Dossie, with which I am pleased to concur, I would like to add to what they have given a few simple ways of discovering negligence or fraud in the manufacture of potash. Skilled potash merchants commonly employ six methods of determining the quality of this product: appearance, taste, odour, dissolution, crystallization and calcination.

1. Whenever they see red potash of a uniform yellow, which they call *gilded*, that colour which the salt extracted from pure beech ash has before calcination, they regard it to be of good quality. The further its colour departs from this tint, the worse is its quality. A few white specks in this potash do not displease them; they are usually nothing but sulphated tartar.

It is important to note that some potashes are a very dark yellow, almost black, without being of bad quality. These are made from fresh ashes, or have not been mixed with others prepared as described above, or the ashes of oak, sweet chestnut or horse chestnut, &c.

2. Good potash should have a bitter taste, burning without any other sensation; however little sea salt it may contain, it can be detected by taste. Bitterness indicates sulphated tartar, when it is strong, soot. Sea salt always reveals a fraud; it has itself been added to the potash or the ashes have been leached with sea water or water from a mineral spring, or the ashes have been mixed with ashes of maritime plants. Sulphated tartar usually promises a good potash and indicates good preparation of the ashes; if the proper taste of the potash is not very strong, or it does not develop at once, this is proof that the potash being tested contains insoluble matter from the ashes, some other earth or some foreign salts.

3. Potash has its own peculiar odour which is not disagreeable. One may regard as bad that which smells of soot, liquid from manure, [or] the residue from soap makers lye which has an almost insupportable smell. Such foreign odours always indicate fraudulent admixtures.

4. Put two or three ounces of red potash into a large footed glass full of pure water; stir well until all the salt has dissolved then stand aside in a cool place for an hour; decant the solution without disturbing the sediment and wash the precipitate three or four times with pure water then examine it. If there is sand it will be felt between the fingers or the teeth; if the precipitate is earth from the ashes it will effervesce with good quality vinegar and dissolve &c. Up to a certain point this matter [from the ashes] is usually proof of negligence and only occasionally of fraud. It would be an excellent thing only to receive potash which gave no sediment.

5. By very slow and carefully conducted crystallization of the first solution, by slight evaporation, appearance and taste can be used to confirm the purity of the fixed alkali salt or whether the potash contains foreign salts, sea salt, or sulphated tartar. The tastes of these salts are easy to distinguish and very different from that of fixed vegetable alkali. When

the latter is exposed to the air in the way indicated it will form prismatic crystals, quadrangular with two pyramids, in the form of a roof, a habit very different from those of sea salt or sulphated tartar.

A small quantity of the latter is not harmful and never indicates fraud. A large quantity is sometimes proof of negligence or ignorance: well water normally containing selenite [$CaSO_4$] in solution may have been used to leach the ashes; it is prudent only to use river water.

6. Red potash which calcines easily in a reverberatory heat from above, with a clear flame, is usually very good. A test using two or three ounces is sufficient to show this. By combining all these methods one can, without doubt, judge the quality of potash and protect oneself against fraud.

MEMOIR ON THE MANUFACTURE OF SHEET GLASS BY THE BOHEMIAN METHOD

Composed in 1775 and sent to the Society of Arts in London

THERE are three principal types of window glass. The first, which appears to be the oldest, is called *crown glass* or *large sheet, large disks,* or *bullion glass* because of the boss in the centre where a mass in the form of a half [plano-convex] lens must be formed for attachment of the punty.

The second type is called *swallow [fish] tail*, because of the form that it has after being flattened, and the third is called *Bohemian sheet, sheet glass* or *cast sheet* although it has been blown.

The first method which is so well known that we can dispense with describing it, produces the least inequality of thickness and least waviness but this advantage is opposed by the defects peculiar to the process and thus inevitable.

1. The glass produced is necessarily badly annealed or *sour* and much more fragile than the other types. At the moment when the workman has fully developed his disk and must detach the punty over a brazier, he cannot put the disk red hot into the *annealing* furnace. Who has not seen that these disks, not being cooled very slowly until completely cold and thus not being well annealed, fail to cut sweetly under the diamond?

2. This method can only produce panes of a limited size, for example 30 × 30 [inches], less often 36 × 30 because, to be so big they would become too thin and that would require the disks to be eight or nine feet in diameter, which is impossible.

The method that produces swallow-tail glass is too imperfect and occasions such great losses in *squaring off* that we need not consider it further.

The third method of making sheets of glass may be divided in two: one produces small *cylinders* and the other large ones. We will only discuss the latter, which is the more interesting, in this memoir.

To leave nothing to be desired on this important subject we next discuss the qualities required of glass to be worked by the Bohemian process; second, we will treat the way of working the glass and blowing it into *cylinders* or *sleeves* of the preferred type and thirdly, the method of opening them out and flattening them without making them lose their surface polish.

1) *Concerning the glass*

Only glass of an average fluidity, the greatest strength, the greatest clarity, as transparent and well refined as possible, can produce the best and most beautiful *sheet* glass. If it is too viscous, like that made only from sand and fixed alkali, it responds with difficulty and unequally to the action of the air from the glass makers lungs; in addition it will be badly refined and unequal in density. If it is too fluid, if too much lime or lead oxide is used, the workman finds it very difficult to control, cannot avoid inequalities in thickness, which are so undesirable, and will take a much longer time to form his cylinder. If the glass is not of the greatest strength ruinous losses will be incurred in transport, cutting, and storage in the warehouse. It will be appreciated that if it has a dominant tint and is not perfectly transparent it will illuminate apartments poorly and give a disagreeable tint to objects seen through the panes.

Defects of refining, *seed* and *blisters, cords*, stones or unmelted sand grains result in a considerable decrease in selling price and cause many breakages. Long experience has proved to us that a good melting furnace and the batch that we are just about to give combine all the qualities that we have just indicated. It comprises 200 lb of very white sand or completely white calcined flint finely crushed and free from reddish specks; 120 lb of the best white potash, well purified; 14 lb of white slaked lime, well ground; and 2 ounces of manganese, if it is of the best quality, a little more if it is of poorer quality.

It is of the greatest importance to mix the different batch materials as carefully and accurately as possible. I have also observed that it is useful to charge it into the furnace two or three days after mixing. The humidity of the air, with which the fixed alkali reacts, contributes effectively to the purification of the glass; we have given the reason in our observations on the art of earthen ware.

We do not believe that *red lead* should be used in the batch; it makes the glass too heavy; the burden on the workman is already very considerable without increasing it further.

2) *Concerning the cylinders*

Before describing the working by the glass blower it is appropriate to say a few words about what is required.

First of all the working hole must be positioned so that the last *reheatings* of the glass maker are not obstructed, not to cause a mishaps such as pock marks [from the glass touching the furnace structure] whilst he is removing the cylinder quickly from the furnace and also rotating the blow pipe in his hands. For large cylinders the working hole should be 15 or 16 inches in diameter, its walls be as thin as possible and it should be set square with the walls and perpendicular to the floor.

It is no less necessary that the blower should have a solid platform built from planks and raised about 18 in above the floor to be able to elongate and open his cylinder whilst holding the blow pipe vertically; that he should have at his right hand a small square wooden trough full of cold water and carrying a small iron fork about six inches high to cool his blow pipe and to support it whilst cooling the different gathers of glass; that he should have to hand a solid block of green wood (beech is preferred) in which many pear-shaped hollows have been made, so that the different gathers can be rounded and shaped, using water; this block is called a *marver*.* He must also have many blow pipes of appropriate lengths and diameters each having a handle of wood 15 to 18 inches long

---

* Not the usual meaning of this term; usually called a *block*.

and 4 or 4½ inches diameter kept tight at one end by a ferrule of copper and at the other by a mouthpiece of the same metal for the convenience of the blower, and that he has many receptacles, fired clay pots 16 to 18 inches diameter and 20 to 24 inches high, in which the cylinders may be stood to cool either in the annealing furnace or outside it.

To have the largest and best-made sheet that is possible, the length of cylinder must form the breadth of the sheet and its circumference the length when it is flattened; that the cylinder must be of uniform diameter along its whole length and the end cap be very short.

Besides the glass being well refined and of good colour, it should be skimmed; this precaution is very wise. If there are *stones* or yellow or red veins, they are usually on the surface of the melt; skimming removes them.

Everything necessary being to hand the boy *gathers* the first lump of glass; when it has cooled sufficiently in the fork over the trough he makes the second gather then begins to *marver* it, to shape the glass in the hollows in the green wood block; when this has reached an appropriate consistency he makes a third gather, marvers it into the shape of a pear in a slightly larger hollow in the block, *pierces* his glass by blowing down the blow pipe to make a cavity three to four inches long and, at this moment passes the pipe to the master glass maker.

Having cooled the pipe and the mass of glass being cherry red, the glass maker makes the fourth gather, marvers it in the largest hollow in the block and blows whilst marvering, he blows holding it in the air perpendicular to his mouth, having his head thrown back, to form the first diameter of the cylinder as near to the blowpipe as possible. This first circle soon acquires a viscosity which no longer allows it to give way to the action of the workman's blowing. In this state the master reheats his glass, rests it on the wooden marver and by pressing on the end and continually turning the pipe, blows and enlarges the first circle; he elongates the glass holding it perpendicular to the mouth with his head lowered; if the cylinder is of equal diameter over its entire length, except for the end, and if the end is sufficiently thin, he casts a button of glass onto the end of this conical part, reheats the glass, and closes the end of the pipe with his finger after blowing into it. The air expands because of the excessive heat of the furnace and acts with a considerable explosion on the part which offers the least resistance, namely where the glass was cast on; then the blower takes his glass from the heat cuts the *edges* with his scissors and, having slightly expanded the opening with tongs held more or less open, he again heats it briskly. Having removed the glass cylinder from the heat he hangs it vertically downwards, turning the pipe rapidly in his hands, and the cylinder opens by itself, the bottom end becoming the same diameter as the rest. If it is somewhat flared he reshapes it over the fork on the trough to the appropriate diameter with his open tongs. After the boy has inserted a dry heated stick into the cylinder the master glass maker detaches the blow pipe from the cylinder with a drop of water and a tap from a small dry iron. The cylinder is carried to the arch on the stick [and put] in a receptacle which is stoppered; it is not removed from the arch, [still] tightly stoppered, until another cylinder has been made. It can be seen that it is important not to remove the cylinders from their receptacles or to carry them to the warehouse until they are completely cold.

The last operations by the master glass maker are very delicate; they require the greatest dexterity and a very precise judgment by eye. I have seen sheets up to 48 inches long by 42 wide made in this way but much more often 42 × 36 inches; even the latter requires a very skilled workman.

3) *Flattening*

First the cap is cut off the cylinder. This is done by rotating the end of the cylinder with the cap over an iron bar, one end of which has been curved into a semi-circle, this part having been heated white hot, then allowing a drop of water to fall onto the heated line of the cylinder. The cap is easily detached by this means.

Before splitting the cylinder it is examined with great care. A large iron rod heated until white hot is then passed inside in a straight line along the whole length of the cylinder under the part that seems most defective. By allowing a drop of water to fall on it or by touching one end of the heated line with the tip of a blade quenched in water, the cylinder cracks along a straight line.

Economy requires that flattening should only commence when one has two or three hundred cylinders; more or fewer may be dealt with at each flattening session in proportion to the volume of sheets. The flattening furnace, Fig. 6, being almost ready, almost hot enough to soften the glass, one takes the cylinder on which one will flatten all the others [the *lagre*], (the largest should be chosen) and places it at the rear end of the passage ACA and makes it slide forward on two iron rods rigidly connected together, it being followed by others, so that they are heated very slowly. When the cylinder reaches the other end the flattener puts the tool inside the cylinder and moves it very carefully into the middle of the flattening slab, briefly reheats it, holding the cylinder [suspended] in the air for a moment so that the edges of the crack do not stick together, then lays it on the flattening slab and makes it open out without wrinkles or folds. When opened out the workman smooths it very carefully with the *polisher*, a piece of wood five inches long by four wide and an inch and a half thick.

The means used to prevent the other sheets sticking to this one during flattening are very simple. A small handful of quick lime or slaked lime passed through a silk screen is thrown into the stoke hole; the finest part of this chalk is carried away in the flame and is deposited on the glass which has already been flattened. If a few grains of refractory or mortar from the crown of the furnace or cinders can be seen on the glass it is necessary to remove them with care. The second cylinder is then flattened like the first; when it is properly flattened the edges are raised slightly using a copper knife fitted onto the end of a long iron rod by means of a sleeve. The fire is allowed to die down somewhat giving the glass a greater viscosity; it is loosened and then pushed by a single tap of the head of the rake onto the second flattening slab in the annealing furnace: as this furnace is much less hot than the flattening furnace the glass there rapidly becomes sufficiently rigid to be stood up and stacked against the iron bars or against the other sheets.

Fig. 6

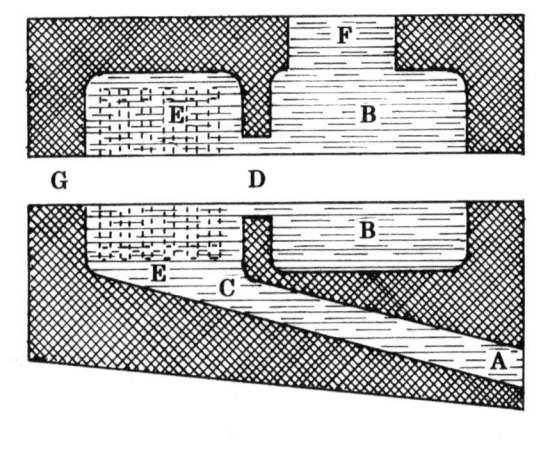

Figure 6. Plan of a furnace for flattening and annealing sheet glass

    AC   Passage for heating the cylinders as they are pushed gradually into the flattening chamber EEE...
    BB   Furnace to receive and anneal the sheets of glass
    DG   Fire box of furnace open at both ends
    EE   Flues to conduct heat from fire box to flattening furnace
    F   Doorway of annealing furnace

When the annealing furnace is sufficiently full or when the flattening is finished all the apertures of both the furnaces are tightly stopped with stone blocks, clay slabs, tiles, and soft brick clay; always beginning at the bottom and finishing at the top: this operation is called *sealing up* the furnace. The next day a little air is let in through the uppermost apertures. The openings are increased every three hours and this is continued until the glass sheets are quite cold.

The sheets are then removed from the furnace and taken to the warehouse. They are squared off with a good diamond and they are rubbed with very finely powdered chalk which improves their polish and makes them brilliant: they can be washed with spoiled beer.

Before this furnace is exposed to a prolonged violent fire it must be constructed to have the greatest possible stability and dried out very slowly.

AC is the passage for steadily heating the cylinders. It should be 20 inches in height to the key of the crown; the floor must be flat from one end to the other so that the workman can easily push the cylinders along with the iron. There must be two iron rods on the floor rigidly connected and sufficiently close together that the cylinders will glide as easily as possible without getting scratched.

G is the *hearth*. It should be 20 inches high to the centre of the crown. During working an aperture only four inches square is left at the end by the flattening chamber so that a stopper can be used when needed.

EE are the flues through which heat is carried from the hearth to the flattening furnace. These flues enter the flattening chamber along each side of the flattening slab, or in the positions marked on the plan.

Over the crown of the hearth, in the middle of the furnace and level with the paving of bricks set on edge, there must be a refractory block four feet long, four inches wide and four inches thick made of four parts clay and five of fine grog. At the end of this block a similar one must be placed in the flattening furnace, also at the same level.

F is the mouth of the annealing furnace. Its height should be in proportion to its width. The front of the flattening chamber, above the hearth, should have a similar mouth which is reduced during working to six inches wide and five inches high.

BB is the floor of the annealing furnace.

D is an arch six inches high in the centre above the refractory blocks to allow the glass sheets to be transferred from the flattening chamber to the annealing chamber. Above this archway there is a hole communicating between the two chambers and one foot in diameter.

The flattening chamber should be three feet high to the centre of the crown from the flattening floor. Two feet from the sides of the floor of the annealing furnace there should be three holes, four inches square, perpendicular to each other and one foot apart; these holes are provided to take one inch square iron bars. These bars are needed to support the flattened sheets, to keep the piles apart, etc. The first holes should be on the floor.

The crown of the annealing furnace should be 3½ feet high in the centre at the side next to the flattening chamber and five feet high at the opposite side.

MEMOIR ON MANUFACTURES USING FIRE
Written in 1775

OF ALL the useful arts those using fire are the most important after agriculture, because almost all of them have an immediate link with our primary needs or because they use raw materials produced in the Kingdom.

These manufactures can be reduced to three classes, pottery, glass making, and metallurgy; dyeing could be made a fourth but it would be a mixture because dyeing may be done either cold or hot. The first comprises tile-making; pottery properly divided into unglazed and glazed; the manufacture of crucibles and furnaces; earthenware of common or fine quality in brown clay and in pipe clay; the different types of porcelain.

The second combines all branches of glassmaking; bottle glass; common and fine window glass; green glass; ordinary and fine quality miscellaneous white glass; crystal; flint; blown and cast plate glass. The third class may be reduced, for our purposes, to iron founding, not all branches of which we possess, and to a few examples of exploiting ores of copper, lead, and antimony.

Are these industries as advantageous to the Kingdom as they ought to be? Have they been carried to the highest degree of perfection of which they are capable? One would have to know very little of them to think so.

The art of tile making is certainly not difficult, nevertheless we have few bricks, floor tiles, or roof tiles of as good quality as they ought to be. There is probably not one furnace for tiles in France which is as economical as it ought to be or as good as may be found among our neighbours. We have an abundance of pure good quality clays yet we always import a large proportion of our crucibles for gold-smithing, minting and chemistry from Germany and Holland.

No one that I know is unaware that the use of glazes on pottery is harmful because almost everyone knows that they are only made from oxides of lead, that lead glass is soluble in acids and fatty materials and that lead solutions, taken internally, are truly poisonous; yet there has been no attempt to replace the lead oxides by materials that would be less harmful. Of all our earthenware factories there are only three, Moustier, Melliona and Rouen which produce wares of an appropriate strength and which will resist heat. A striking proof of the ignorance where this art is found is that the factories of Paris, Rouen, Langres, &c. take sand from Nevers for their enamels although they have just as good sand at hand.

Our fritted porcelains do not deserve the name of porcelain because they cannot withstand a degree of heat much above that of boiling water [and] every change from hot to cold or *vice versa* is harmful to them. Our other types of porcelain are very inferior in both the strength of the body and the gloss of the glaze to those of Japan, China, and Saxony. However, we have in this respect an inestimable advantage over all other nations. In China and Saxony it is necessary to use two different materials to make up the body but since the discovery of the white clay at Saint Thérié near Limoges, we can make it from only one when it is treated appropriately. However numerous the products using this clay might be, you may be sure that this precious resource remains hidden in the ground and that it is used only for small saleable articles and those at an exorbitant price.

Since 1760 there has been a very favourable movement in the art and commerce of glass making in France. Before that time miscellaneous green glass products were as expensive as white flint glass is today and almost two thirds more soda was imported from Alicante and Sicily. This fortunate revolution seems to be entirely due to the works of M. D'Antic, particularly his *Memoir on the Perfection of Glass Making* which gained the prize of the Royal Academy of Sciences.

There is, however, much that remains to be done in this important branch of commerce. Although the manufacture of excellent bottles appears easy we have only two, Sèvres and Folembray in the forest of Coucy, which provide them. The Bohemians retain their advantage over us in miscellaneous white flint ware, large sheet [crown] glass, crystal both cut and not cut. In spite of taxes very considerable quantities are imported into the Kingdom each year; they are found in the shops of nearly all our towns of the first and second rank.

England provides us, at considerable expense, with watch glasses, cut crystal, and optical flint glass. Our blown and cast plate is of poorer quality for transparency, colour and tint than that of England and some German factories; that from Nuremberg has an extensive trade with France because its price is 25% lower that of our own, even for small quantities.

Unfortunately we do not shine in metallurgy. In this respect the Germans have very great advantage over us; however, we have an abundance of different types of ores and, first of all, we need cede nothing to our neighbours in iron ores, the most precious of all: we have rocks of the most favourable type in the Vosges, in Lorraine, Brittany, the Pyrenees, Roussillon, Pays de Foix, the Dauphiné, &c.. In the last four areas we have spathic iron ores and alluvial iron ore in most of the other provinces. Nevertheless we cannot flatter ourselves that we have iron as good, of as good purity, as that of Styria, Carinthia, the Tyrol, Roslagen in Sweden, the Duchy of Brunswick, the Electorate of Hanover, Spain, and North America; nor steel ingots of as good repute as those of Deux Ponts;, or steel plate as good as that of Carinthia and Styria or hardening steel and cast steel as good as those from England.

We are so little advanced in this interesting area that we import every year many millions of iron from Siberia, Sweden, and Spain &c.; different types of steel from Germany and England. Foreign iron wires and plate are far superior to our own (there is not one bucket-maker in the country who does not use Swedish thin sheet-iron). We have not been able to produce a good sewing or embroidery needle and always import them from England. We have not suspected that there could be means of purifying and softening cast iron to the point of making it accessible to boring, cutting, filing, drilling and lathe work. The English have been more fortunate; they have discovered a simple method of purifying cast iron and as a result, their naval cannon are the best in Europe and they alone produce rolls for crushing sugar cane and the vats used to produce and purify this precious substance, even in our own colonies. What is it that gives these proud insular men their extremely high reputation in ironmongery? It is the care that they have taken to use in their case-hardening and bulk steels only iron from Styria and from Dannemora in Sweden.

We have, I believe, two copper ores that are being exploited, one from Saint Bel and the other from the Pyrenees. Neither of these has produced rose copper as pure as that from Sweden and neither has until now produced black copper; is this the fault of the ores? I do not think so.

There are lead ores in various provinces but there is scarcely one in the Kingdom that is being properly exploited. We have not yet produced as pig lead as pure as that which we import from England nor silver of ordinary purity.

Pharmacy, painting of buildings and equipment, glass making, pottery &c. consume enormous quantities of red lead; innumerable attempts have been made without success to make it within the Kingdom yet we continue to pay Holland and England the costs of both the raw materials and of manufacture.

The Bourbonnais, the Auvergne, and Limousin are rich in ores of antimony. Yet we import, through Marseilles, the majority of its preparations. Perhaps we would even import white metal for printing types if chance had not thrown up a factory for this useful operation at Orleans and another at Pontoise.

A rich source of cobalt has been discovered near Baigory in the Pyrenees, but I am told that it has been abandoned; consequently we continue to pay Saxony for the zaffre and various types of cobalt blue that are indispensable to our glass houses, potteries, and enamel painters.

The picture that we have seen of the present state of our industries requiring heat is not to our advantage but it is accurate; all these comments are unfortunately only too true. Inspection of the records of import taxes would at least provide proof that foreign countries sell us every year their manufactures worth immense sums. Each year more than ten million of iron ingots, wire and sheet, and steels or various types are imported from

Sweden, Siberia, Spain, Deux Ponts, Styria and England.

Are there ways of stopping these ruinous importations? Is it possible for our industries using heat to achieve at least the degree of improvement seen among our neighbours? I see no sound reason for answering with a negative; I like to think that we can even raise our industries to a higher level of perfection. Local circumstances are very favourable to us. It is only because of their low price of wood that several European nations take trade from us; but a better administration of our wood and coal would provide us with an ample compensation. There is no industry needing heat in which well prepared coals could not be employed with advantage; the example of Great Britain proves it.

However different from each other these useful arts of which we have spoken may appear to be, they lend each other mutual assistance and share common principles. One cannot improve one of them without having theoretical and practical knowledge of the others. It is impossible to make progress in the art of glass making without knowing the nature, preparation, composition and use of refractory clays and it is pottery that provides these insights. One could not take a single useful step in metallurgy if one did not have proper ideas about refractory clays, fluxes, natural absorbents and agents of vitrification: pottery and glass making provide us with these ideas.

There is more, to improve a useful art it is necessary to become familiar with the principles peculiar to it, to know the steps in its operations in the greatest possible detail and to have the talent that providence or a very lengthy experience may give us successfully to apply theory to practice.

After that might we expect perfection in the useful arts from our manufacturers? To do so would be misleading because the majority do not possess the necessary insight. What we can hope in this regard for the future is only too clearly indicated by the past. If a manufacturer makes a discovery he retains it for his own advantage and his interest is not always the same as that of the State.

At a first glance, books which discuss the useful arts appear to be very helpful but they have not until now produced any large effect. The majority, one can even say, have served only to perpetuate prejudices and errors; the reason is easy to understand. Nearly all those who have written about our industries using heat have not understood their practice in sufficient detail. Seduced by the misleading observations made in their laboratories they always draw conclusions about the large scale based on the small, a method of reasoning almost as defective as that which draws a general conclusion from the particular. To avoid being misled, however simple a manufacturing process may seem to be, it is necessary to know all the details of its operations for many years and this knowledge is indispensably necessary to be able to contribute to the improvement of the process.

On the other hand one may state without fear of being unjust that the great proportion of our manufacturers are in no position to profit from the teachings of our writers; one example will prove what we have said. We have a translation of the treatise on iron by Swedenborg; the *Art of Converting Wrought Iron into Steel and of Softening Cast Iron* by the celebrated Réaumur; a good description of the processes used for almost all the iron-making of Europe, *On Fire* by Jars; the *Art of Foundries* by the Royal Academy of Sciences; many excellent memoirs on this precious art from Aristotle to our own century. What has been produced by the works of these knowledge authors? Almost nothing contributing to the improvement of iron manufacture. The master founders have not been able to profit from them and our iron products have not made any progress. The irons are even worse since workmen have been taken on by the thousand. A hundred attempts have been made to make steel on a large scale but all these entrepreneurs have been ruined; the works of Neronville and Soupe has finally succeeded, after spending more than two hundred thousand pounds on unsuccessful trials, by using, like the English, iron from Roslagen in Sweden. However this demonstrates again how we always import iron and steel from foreign countries at enormous cost.

To believe that our industries using fire can be improved without the co-operation of the Government, seems to me a dangerous illusion. We must not allow ourselves to be misled by the example of Great Britain where the people concerned are rich and all animated by patriotic pride;

unfortunately it is not the same in France. One must not however think that the industries in England are entirely left to themselves; the Government rewards important discoveries with exemption from taxes and by an exclusive patent for forty years. Also a society of interested gentlemen distributes one hundred thousand pounds annually to encourage the arts and commerce in that Kingdom. Our industries can only prosper and make rapid progress under the informed protection and the invigorating hand of the Sovereign.

But what might be the most advantageous influence of the Government on our industries using fire? Might it be to come to their aid with large sums of money and allowances for the entrepreneurs? These methods have often been tried but I do not believe that they have ever had any marked success.

To attain the highest degree of perfection our industries need only solidly based instruction, instruction where the theory always illuminates practice and is itself constantly justified by large scale experience; they can only expect such aid from the Government. It seems to me that the Government could bring about this desirable revolution in two ways.

1. They could charge people of insight who have long left the industries, under whatever title they found appropriate, to visit each year a certain number of these establishments. With them they could make an exact accounting, instruct the proprietors in everything that might be of interest to their own enterprises, to make their products more perfect and produce a greater output. These instructions, always related to local circumstances and given on the spot, which would speak to the eye as well as to the understanding, would surely not fail to produce the desired effect. However, we must recognise that by this means alone improvements could only come step by step.

2. The Government has another no less efficacious means of attaining the most rapid and most extended progress. This would be to establish one or more schools, where the principal operations of the industries using fire would be carried out on a large scale, in which the practice would always be illuminated by the torch of a theory based on reason and

where one would only encounter as true principles those soundly based on large scale experience. This school would, above all if the zeal of the teacher and his pupils was sustained by requiring a public examination at the end of each course, certainly have the desired effect.

Such an establishment, dignified by the wisdom of the Sovereign who governs us and to the zealous insight of his ministers, run with economy, would not be an object of great expense. If it were established in a German type of glass house it would cost no more than twelve or fifteen thousand pounds a year. By adding to the furnaces of the glass house two other furnaces [one described as *à courbe* and the other as *à lunettes*: neither term is explained] and an English reverberatory furnace, one could there carry out all principal operations of the industries requiring heat.

These two alternatives are not incompatible and it is to be hoped that the Government would decide to unite them: the same person could supervise the production during the fine weather and give his theoretical and practical courses during the winter.

This is how our industries needing heat could rapidly change their aspect, how they would necessarily make very rapid and very considerable progress, how we could cease the imports that drain us of immense sums of money, how we would be able to supply, to advantage, products from our factories to foreign countries and how we could open up a new source of riches for the state and for its subjects &c. &c.

It is not easy to understand the reasons why these precious industries have until now been left to themselves without assistance and without aspirations. It is much more difficult when one considers with what attention and munificence other industries have been treated! The institutions provided for them are immense; one of the most obvious is to have given them five inspectors general and individual inspectors for each activity, even in each manufacturing town.

From their beginnings the inspectors appear to have had three functions to fulfil: they were envisaged as teachers who could instruct the manufacturers in their early development, they could make regulations for them which would themselves represent the best practices of the time, and they could report to the Government on their conditions and their progress.

It is apparent that one should not understand these industries, principally those of cambric, silk, gold, and silver to have been carried to the peak of perfection, to have no further need of assistance and to be able to fly on their own wings. Also it is clear that the special inspectors have long since ceased to be teachers; they no longer have instructions to give to the industries; their functions are restricted to accounting for the range, measuring the widths and lengths of the pieces, judging the good and poor tints and reporting to the Government on the condition (nearly always the same) of the industries. The jurymen of the communities could equally well fulfil the first task and at no expense, and the offices of the intendants could do the second. It is evident that at least the provincial inspectors are perfectly useless. With how much less than they cost the state annually could one bring the industries using fire to perfection.

This would not be the only advantage gained from such schools; they would be the source of many others related to mineralogy, the manufacture preparation and use of salts, agriculture, the preparation of colours for dyeing, painting, thread printing on silk, cotton, &c.

OBSERVATIONS ON THE ART OF ASSAYING ORES BY HEAT
Composed in 1774

Presented and read to the Royal Society in London in 1774

THE assaying of ores is an interesting part of chemistry from which we may expect the improvement of mineralogy, precious assistance for the arts, precision in metallurgical operations, important resources for the State, and so on. These advantages are too real and too far reaching for docimasy not to have received the attention of all chemists.

There is no other part of true physics that has been so constantly and carefully cultivated but has it had success proportional to the number of people occupied with it or to the time and effort expended?

Although it seems rather paradoxical that the answer should be "No", I am convinced that friends of truth and those properly trained will agree with my opinion. The only favour that I ask of them is to do me the honour of following me, verifying accurately my sources and my experiments and allowing me an indulgence in proportion to the difficulty and the importance of the undertaking.

The early chemists mentioned docimasy and treated the subject. They prescribed the separation of the ore as well as possible from all other heterogeneous materials; taking a representative sample; grinding it; roasting certain types; weighing exactly the sample to be assayed, mixing it with a certain number of parts of borax, fixed alkali salt, white flux [calcined tartar with half as much nitre], black flux [calcined tartar with an equal amount of nitre], *sal ammoniac*, salt for glass making, powdered carbon, glass, &c.; covering all with rock salt, heating it in a good crucible under a muffle at a convenient heat, and so on.

This art appeared so simple to their successors, so well considered and detailed, the means so in agreement with all chemical knowledge, that their disciples felt obliged to copy them, as in almost all the useful arts. We can agree that the most impressive appearances appealed to them.

To separate the metals from their mineralizers, sulphur and arsenic and from siliceous gangue, lime, clay, &c. which cannot be done by washing, it is necessary to open or divide the ores by melting them. Does not everyday experience confirm that the salts mentioned above are the most active fluxes? Also that, for a long time, they have been regarded as the only ones?

To make the assay exact, to extract all the metal that it contains from the ore, it is necessary to ensure that no particle of metal has lost its phlogiston and that it be given back to any particles in the ore that have lost it. Can the means indicated achieve both these aims?

Might those who have published treatises on assaying, such as Schelinder, Schlutter, Cramer, Ercker, Kiesling, Geller, have added much to what had been said by earlier chemists? Almost nothing new is to be found in their works. They have collected together scattered precepts but without giving them any true development; they have united them in one point of view but without deducing any illuminating and clearly correct principle. The methods that they have taken the trouble to use to bring them together are so imperfect that nearly all the nuances, the differences, that may noted in each type of ore escape them. The proof that this is no exaggeration is that when two chemists assay an ore by the methods of these masters, they constantly obtain different results. However slight these differences may be, they deserve the greatest attention because they may lead to considerably larger differences in large scale operations.

This is not the only proof of the inadequacy of their works and the imperfection of their art. Examination of their doctrine, discussion of their procedures and methods will provide us with many others.

Docimasy is not an art merely of pure curiosity. Its aim is to determine how to exploit ores, to establish the ease of doing so, and to fix the advantages; it must achieve on a small scale what metallurgy does on a large scale, extracting from the ore all the metal that it contains and doing so with the least possible expense.

Who can believe that these two arts, which in truth are only one, should necessarily use the same methods to achieve the same results? Are docimasy and metallurgy the same in this regard? Should they be? Could an iron-master or a copper-founder be expected to bear the expense of using borax, fixed alkali salts, sal ammoniac, or even rock salt, when working on a large scale? It is no bad thing that these materials cannot be used in large scale operations because there would be other disadvantages. These are new reasons demonstrating the imperfections of assaying ores.

These salts are so fluid on melting that they would run to the bottoms of the crucibles; they would also corrode the vessel or tank before the metal, having been separated from its ore, could displace them because of its greater specific gravity.

In assays where it is possible to grind samples to the finest powder and to make mechanical mixtures as perfectly as desired, these effects are scarcely noticeable. However, there are other salts which have not been recognized as being no less certain and no more opposed to the purpose proposed.

It is known that sulphur and arsenic, with the favour of an open fire, are good ways of reducing imperfect metals to oxides. It is also known that this semi-metal and this mineral only have an affinity for neutral salts because of their phlogiston and, in the case of decomposing sulphur, by its acid formed with neutralized salts, that is by a weaker acid.

Neutral salts which contain acid of vitriol [SO_3] and acid of rock salt [HCl] are thus with good reason included among those useful for calcining metals. It is no less certain that these salts can contribute to the vitrification of stony [siliceous] or earthy gangue in ores[a] and that this vitrification is absolutely necessary for the separation of the metal. A multitude of observations leaves no doubt that these neutral salts, at the heat used for glass making, have no great affinity for phlogiston nor that they acquire, at the

---

a. See the memoir of Pott on glass gall and my memoir on the causes of bubbles.

same degree of heat, sufficient to make them volatilize.[b]

What use can these neutral salts be in docimasy according to these incontestable facts? They can produce the most pernicious effects and make fail the purpose for which they have been proposed; they can deprive the metal of part of its phlogiston, to produce slag; even to remove part of the inflammable principle that has been added in the powdered carbon, black flux, or *brasque* [a mixture of clay and carbon used to coat crucibles].

If any doubt remains in this regard it may be removed by a very simple test. Have two baths of rose-copper, made as perfect and equal as possible, from which all dross has been removed, kept at the same heat. Then pour over one of these baths sufficient molten rock salt or gall to cover it and you will see, 1) that the salt becomes coloured, which can only be at the expense of the phlogiston in the copper, 2) that this salt volatilizes and 3) that the copper is covered by considerably more dross where the salt was used than is found on the one simply exposed to the heat. It will also be found that the former has lost more weight than the latter. But, one may say, if the gall, sulphated tartar, rock salt, or kitchen salt do not form a glass with the gangue of the ore, do they not at least protect the metal from the destructive action of the flame? As a pool floating on the metal do they not prevent the loss of phlogiston?

This is a pure illusion! From the moment of everything having melted these salts float on top, so can one conclude that they are an obstacle to the calcination of the metal, is there not always some dross between the metal and the salts? One would need never to have cast one's eye on experiments made with care to ignore this. To be convinced, would it not suffice to know that these salts cannot enter into combination with either the metal or its *matrix*, that these three materials have different specific gravities and that the salts are the lightest of the three.

Lastly, to leave no doubt about the effect attributed without foundation

---

b. See my memoirs on the perfection of glass making, faience. and the electrical fluid.

to neutral salts, let us examine what happens during the operation. These salts are the first to melt and as they melt they find their way to the bottom of the crucible, flowing around the particles of metal and carrying away part of their phlogiston; as soon as the salt has all flowed to the bottom, the metal is exposed and the action of the flame takes away some of its phlogiston, the metal begins to soften, its particles coagulate, and form a mass of greater specific gravity. The salts are thus forced to rise to the surface again and remove more of the phlogiston from the metal particles that they contact. The metallic earth that is found in the ore, or formed by the action of either salt or flame, vitrifies to include the gangue present, then the metal begins to melt and its droplets, having the greatest specific gravity, fall to the bottom of the crucible and coalesce. It is evident that it is the glassy slag produced from the metallic part and not the salts that protects the metal from the action of the flame.

I will not discuss the use of sal ammoniac in docimasy which leaves me speechless and appears ridiculous. It has no good effect and is dissipated by the time that the mixture has become no more than merely red hot.

However, can fixed alkali salts have effects as desirable as those of the neutral salts are pernicious? This has been believed because insufficient attention has been paid to their properties and without considering what happens during the operation. All the fixed alkalis, however charged they may be with phlogiston, lose it almost completely before melting. Thus one cannot expect a greater effect from *black flux* than from *white flux* or from *red potash* than from *white potash*.

The fixed alkalis are much more avid for phlogiston than the neutral salts and they only yield it to the latter when they change state and become a constituent part of a glass.

Réaumur and Sage have observed that malachite produces more copper with powdered carbon than with *black flux*.[a]

Alkali salts charged with a certain amount of the inflammable principle

---

a. See pages 208–209 of the *Chemical memoirs* of Sage.

do not, at the heat of melting, become less volatile than the neutral salts. This is a fact known in Chemistry but neglected in docimasy.

These observations suffice to show that the operation with fixed alkalis occurs just as with neutral salts, except for very slight differences, as anyone can easily confirm. It may be said that the fixed alkalis have a greater and indeed complete affinity for the mineralizers sulphur and arsenic and that they show to the greatest degree the property of combining with the gangue of ores to form glass. I entirely agree. Whether sulphur is present in an ore as a mineralizer or merely as itself, fixed alkali salt removes it completely and forms a sulphurous paste but one cannot ignore the fact that this paste dissolves metals and that, at the heat of melting, it takes away their phlogiston, or that it will volatilize completely if not promptly decomposed and converted into sulphated tartar; thus we are brought back to the neutral salts.

Arsenic loses its volatility when combined with fixed alkali salt but it is no less certain that, by this combination, the alkali loses its fixedness at the heat of melting and becomes volatile which makes it incapable of entering into vitrification with the matrix of the ore. In that case how can the phlogiston be conserved in or returned to the metal? There is more. Sulphur and arsenic are not the only mineralizers. It had been believed that there were no others until the chemists discovered that sulphuric and hydrochloric acids in silver, lead and mercury waste, aerial acid in spathic ores of iron and lead, and volatile alkali also act as such.[a]

One may think that fixed alkali salt would be no use for scrap metals. Mineralizers would be converted in rock salt or *salt of Sylvius* [presumably some salt formed from pine resin], sulphated tartar, or Glauber's salt, and would lose their fixedness by combination with aerial acid or just by contact with aerial acid.

Many people believe that the only means of extracting the metal from

---

a. See the *Mineralogical and chemical memoirs* of Sage. Page 119 of the mineralogy of Cronstedt, French translation, and the Experiments made in order to ascertain the nature of mineral substances by P. Woulfe FRS.

an ore is to vitrify the matrix but a considerable proportion of ores never behave like that. Where is the gangue of copper, arsenic, or iron pyrites, of some *falherz* [ores of arsenic and antimony], of malachites, haematites, most spathic iron ores, of many galenas, ores of white, green or red leads? It would be absurd to consider the small proportion of absorbent earth or sulphurous paste that the knowledgable Sage discovered in some of these ores as a matrix.

The glassy slag often formed on the *bottom* [culot] is often a surprise; it is largely, sometimes entirely, formed at the expense of the metal.

We add only one more reflection on the use of fixed alkali salts in docimasy; this is that they are particularly pernicious for ores with siliceous gangue. Fixed alkali combines with siliceous earths and flinty rocks to form glasses of the most viscous type that is known. These never rise to the surface completely and always retain some of the metal: this is a fact that cannot be doubted.

It seems that one cannot disagree that at least powdered carbon is no better indicated. Who will not believe that the action of the phlogiston is completely dissipated by the action of the fire before the metallic earths are ready to receive it. Amongst other things, how is the quantity of free phlogiston, like that of the carbon, that a given quantity of an ore requires to be determined? However, if a metal is provided with an excess it will definitely be exposed to a loss; a quantity will volatilize which will be in proportion to the excess of the inflammable principle supplied.

The sublimation of metals in the manner of Glauber and the complete volatilization of a given quantity of lead by means of phlogiston[a] provide striking proofs to be convinced without further experiments. One has only to examine the material that hangs on the walls of foundry chimneys where only charcoal is burned, particularly small foundries where a charcoal fire is used to extract lead from lead-workers' ashes; considerable quantities of metallic substances will be found there.

It should not be surprising that the art of assaying has made so little

---

a. See the memoir of Geoffroi in those of the Royal Academy of Science for 1753.

progress: it has not been on a sound basis and what might lead towards that has been neglected.

I have found nothing concerned especially with the question of what is the melting of metals. Are there differences between this and the melting of glasses or salts and what may be the causes of these differences? If we could be so fortunate as to determine the former and discover the latter, docimasy would have fruitful true principles and procedures in which no error was to be feared.

I do not underestimate the difficulty of these researches but the matter is too important not to merit our efforts; if conducted by the light of experience and observation we need not fear being led astray.

We know only what the salts, above all the fixed alkali salts, vitrifiable materials and metals require for their fusion and that, when maintained in a violent fire, each of these fusions shows its own peculiarities. However, they may have the same principle which may produce different effects only because of the nature and proportions of the bodies in which it is combined.

1. Molten salts are more tenuous than either molten metals or glasses. Their fluidity approaches that of water whilst the others are far from that condition in the order that we shall give. The molten salts also penetrate the crucibles much more easily. Proof that this is not only because of their ability to form a glass with the material of the crucible but also that the least attacked crucibles are those that are well compacted and that they last longest when care has been taken to coat the interior with an ordinary white glass to seal the pores.

Metals acquire a little less fluidity, retaining rather more tenacity in their fusion. The pores in the crucibles would have to be larger for the metal to escape through them, although it must be remembered that the weight of the column trying to enter them is proportionally considerably greater.

Molten glass is much more tenacious than the salts or the metals; it is even pasty and capable of being drawn into threads. But all glasses are not equally viscous; those made [only] from flint-like rock [silica] and fixed alkali are the most viscous, those to which metallic earths, lime, fusible spars, brown or black basalts, are added are the least; the viscosity decreasing as more is added.

2. If you add powdered carbon or other materials capable of expansion (or of increasing that effect in others) to molten salts, metals or glasses, different phenomena occur. In molten salts there is rapid bubbling of short duration in the bulk of the melt, the salt becomes coloured and part of it volatilizes. In metals there is no obvious effect except a whitish vapour which rises above the melt without any obvious vigorous motion.

In glasses these phlogistic materials cause a considerable swelling, sufficient to overflow the rim of the crucible but settling by degrees and producing large bubbles which burst with a kind of whistle; the glass becomes coloured yellow and a whitish vapour is formed but is much decreased as the vitrification becomes more perfect.

What we have observed suffices to prove that phlogiston is a constituent of metals but never of glasses.

3. The intensity of the fire is alone sufficient to give a perfect metallic fusion. However, the long continued action of the fire is as necessary as the intensity of heat in producing a good glass.

4. There is another more important difference to be considered. This concerns the different effects of the duration of the fire on the fusion and complete purification of salts, metals, and glasses. Salts lose an increasing part of their volume and weight if heating is prolonged, metals become oxidized, able to form glasses, decreasing their specific gravity whilst increasing their absolute weight. Only glasses suffer only negligible changes. I have held fine white glass at melting temperature for two months without discerning any decrease in volume or weight. One here sees a remarkable thing, that long continued heating can convert metals into glasses but never convert glasses into metals.

5. Molten metals do not adhere to the walls of the crucible but glasses do and indeed adhere very firmly.

6. On cooling metals take on a convex surface but glasses do not.

7. In the open air metals solidify much more rapidly than glasses.

However numerous these differences may be and how clearly defined they may appear to be, they may be due to the same cause modified in different ways and we should forget nothing in seeking to confirm this. This important discovery seems to me result from the scrupulous examination of a certain number of facts. By approaching them we can find what they truly have in common, namely the cause that we seek.

It can be agreed that, by dissipation of their phlogiston during melting, metals lose their colour, their brilliance, their malleability, and their specific gravity. I do not believe that there can be any doubt about that. It is no less certain that on conversion into oxides they increase in absolute weight and acquire the singular property of being able to form glasses by themselves. Are these the results of one cause or two? It seems to me very likely that there is just one cause when the two effects occur simultaneously.

The majority of physicists and chemists have only considered the increase in absolute weight. Others have attributed it to *air* or *water*; others have assumed it due to an *increase in volume*, some to *particles of fire*, yet others to *saline materials present in the atmosphere*. In his knowledgeable treatise on phlogiston Morveau attributes it to the *simple absence of the inflammable principle*; Meyer and Gellert to *an acid* which Sage called *phosphoric acid* and he attributed to it, without any proof, the ability of metal oxides themselves to form glasses. The two phenomena are, we may claim, inexplicable if we attribute them to two separate causes. Reduction of metal oxides alone is sufficient to show that this assertion is not guesswork. In effect, by giving them back their metallic properties by means only of phlogiston, by either the dry or the wet route, they lose equally their increase in weight and their ability to form glasses; but we must not neglect any means of showing this truth in the clearest light.[a]

---

a. See the notes to our *Memoir that won the prize*, where many proofs are given.

The volatile spirit of sal ammoniac made with metal oxides is no different from that formed with ordinary chalk: this operation offers many phenomena that have never even been suspected. It appears that only the property of this spirit to fail to effervesce with acids has been seen. The hint that this singular property gives about the consequences that naturally follow has not been taken; that is that metal oxides, like ordinary lime, remove aerial acid from the volatile alkali of sal ammoniac. Two observable changes to metallic oxides caused by this operation have not been noted: an increase in weight and a greater disposition to vitrify. Can one doubt that these two simultaneous changes are due to one and the same cause? But we have not yet exhausted the proofs of this truth.

Réaumur was the first to observe that cementation of glass with lime converted it to porcelain and made it lose its colour, its transparency, and even its ability to melt again or even soften in the most violent heat. This knowledgeable academician, like Pott after him, thought that this conversion was entirely due to the finest particles [atoms] of the lime penetrating into glass during cementation. If this were true the glass would gain weight and the lime lose it. However, I have frequently demonstrated the opposite, the loss in weight of the glass being considerable and increasing as the glass is of better quality and transparency. [A dubious claim]. It is no less certain that cementation of glass with silica sand achieves the same result just as readily as when lime or plaster is used.

The open pores that Réaumur observed on the surface of his porcelain should at least have made him suspect that the glass had lost weight during the operation. They would appear to be an indicator that some of the material has been turned into vapour by the heat and has managed to escape its prison.

This porcelain does not retain any other property of the glass but also resists the highest degree of heat and requires about a quarter more flux to form a glass than does the equivalent weight of silica. Has the glass then lost one of its constituent parts by cementation? It may be found [instead] in the fixed alkalis and in the metallic oxides.

This common principle or agent, which appears to us to be the only principle involved in forming glass, cannot be phlogiston because that can be removed from a glass without it ceasing to be glass. We believe we have proved that to be so in our observations on the making of faience, on the electrical fluid, and in the notes to our prize-winning essay. It is no more difficult to extract it from the fixed alkalis without them losing their alkaline properties. Metallic oxides could not owe their ability to form glasses to phlogiston because they only acquire the former by losing the latter.

For a long time I have been convinced that the vitrifying principle is an acid salt, however, I have been unable to find it among our three mineral acids, which forces me to propose, after a large number of experiments that it must have a very considerable specific gravity and a high degree of fixity, &c. The indefatigable Meyer and the knowledgable Gellert have reinforced my belief. I agree with the former that the *acidum pingue* of the ancient chemists, too neglected by the modern ones, is the cause of the double phenomenon.

In 1770 Scheele discovered in fusible spar [calcium fluoride] an acid [hydrofluoric] that Marggraf had recognized and which seems to me to have all the characteristics of *acidum pingue*, Sage named this acid *phosphoric acid*. We will prove in the rest of our works that the most suitable name for this is phosphoric acid and that different forms of this acid play a very important part in the three kingdoms of nature. [Hydrofluoric and phosphoric acids are confused in this discussion; most of the comments apply to the latter].

This acid has properties too singular for it to be confused with any of our three mineral acids. It has no smell and is colourless and can be put on the tongue without any danger; it is heavier than even the most concentrated sulphuric acid; it is the most fixed; when united with an appropriate base it resists heat and and vitrifies; it has so great an affinity for phlogiston that it easily removes phlogiston from other acids and by that combination it becomes volatile, more or less elastic and able to decompose glass, &c.

The phosphoric acid so developed appears to me to provide the reasons for the three different types of fusion and of the phenomena that we have indicated which seem incapable of explanation by any other means.

It also provides solutions to the many problems of which we have already spoken. Why are semi-metals not malleable? Why, on the whole, are they volatile at a certain degree of heat? It is because they conceal a certain quantity of the phosphoric acid, variously modified by the inflammable principle. In 1772 the knowledgeable Lassonne demonstrated this for zinc in a memoir that contained many new opinions, ingenious experiments, and profound researches. Why are the noble metals incapable of being calcined by ordinary means; why are they the most ductile? It is because they contain the least possible amount of phosphoric acid; perhaps they do not contain anything else that can be a constituent of phlogiston.

Although the common metals are more or less malleable, more or less ductile, do they have greater or lesser affinity for the electrical fluid? It is because the phosphoric acid, modified in different ways by the inflammable principle, enters into their composition. I am persuaded and hope to prove that most of the differences that may be found in the quality of irons have no other cause. Those that are the effects of poor fabrication or insufficient care in the dissipation of zinc in founding, or under the hammer, are not as numerous as is commonly believed.

We are, I believe, ready to reply in a precise manner to a question that is no stranger to the art that we are discussing. Can one make glass that is malleable? We have been assured that it was demonstrated to the Emperor Tiberius and to Cardinal Richlieu. Malleable glass is nothing but a chimera which would rely on a number of impossibilities; that which removes the metallic properties from metal oxides makes them approach the properties of glasses and *vice versa*. I believe that it is clearly proved that phlogiston is a constituent part of metals but that phosphoric acid is found in them only by accident; although that acid is a constituent of glasses, yet phlogiston found in glasses is not one of their constituents.

What method should replace that of the ancient chemists and what procedures should supplement those that they gave? What materials should be used instead of those that they indicated? These are the tasks that we have set ourselves and which must now be fulfilled.

We will not dwell on the preparation of the ores. We will only note that calcination is too widely used; it is only necessary for ores mineralized by sulphur or arsenic, or containing appreciable amounts of one or both of them.

We only consider the materials commonly called fluxes that are employed in assaying. From what has been said above these fluxes should have six properties.

1. The greatest degree of fixity at the heat used for melting, so that they can combine with the matrix of the ore to form a glass and protect the metal from the action of the flame.
2. They must melt speedily so that they permit only the minimum possible loss of phlogiston from the particles of metal.
3. They must have the greatest possible affinity for the *gangue* so that they convert it as rapidly as possible into glass.
4. The molten glass should be as fluid as possible to allow complete separation of the metal to the bottom of the crucible.
5. The fluxes should have as great as possible a capacity for the mineralizers so that they can remove them from the metal.
6. It is no less necessary that they should contain sufficient phlogiston that they do not lose it solely by the action of the fire and so that they can transfer it to the metal oxides.

No one can be unaware that different types of earth or rocks have long been used successfully as fluxes in metallurgy. Iron-rich clays and limestone are widely used in iron-smelting under the name "*castine*"; brick clay is added to iron ores already containing limestone under the name "*herbue*". Fusible spar and schist are used in many copper foundries, and so on.

Although metallurgy has not given the reasons for the desirable effects of these mixtures, it can be accepted that, in this respect, they have made more progress than has been seen in docimasy. Chance has led them to the useful procedures that they employ, for which we should be grateful. There are those who perhaps have attributed to the knowledgeable Pott the great truth that each type of mineral earth – like gypsum, or clay, or silica – that is not capable of forming glass by itself can do so when mixed with others in suitable proportions; Pott set out to demonstrate this clearly by innumerable experiments.*

His *Lithogeognosie* offers us many fluxes suitable for use in the assaying of ores. We will examine a few of them, adding the observations that long experience permits us to make.

Pott was the first to find that the most refractory lime combined with clay forms glass more quickly as the proportions approach what he recommended and as the clay becomes less pure. He found that almost the same phenomena[a] were produced by mixtures of clay and gypsum or fusible spar and lime, the last giving the quickest melting.

The blackish, brown, or dark grey basalt of the Auvergne is more fusible than this mixture but once melted seems to me not to attack clay, or earths of lime or silica with as much energy; its glass retains more phlogiston but is not so fluid as that given by the mixtures mentioned above. It is appropriate for reducing metallic oxides and it does not corrode the crucibles so vigorously. The last effect is probably due to the silica that basalt contains.

These are the materials that we should use instead of those indicated by the old assayers but, to make these more certain and more convenient in use they should be converted to glass before use.

Glass is by far the best material for reducing metallic oxides, releasing them from their ores and removing the mineralizers from them when it contains the most colouring principle, phlogiston, and its composition contains a lot of lime.[b]

Those who use lead oxide in their glass find much more of this metal in the bottoms of their crucibles as their composition contains more colouring principle or glass-makers salt. It is known that neutral salts

---

* See his *Lithogeognesie*.
a. See chapters 1, 2 & 3 of his *Lithogeognosie* (French Translation).
b. See the *Mineralogy* of Cronsted, p. 16, French translation.

remove from the glass the phlogiston which colours it[c] and thereby prevents reduction of the metallic oxides.

One should not ignore the fact that some ferruginous pyrites is found in some grey refractory clays, notably those of Suzi in Picardy. The grains of pyrites left by the carelessness of the workers in crucibles made from this clay give up their metallic part in crystal glass and form a brilliant round, malleable button of metal, which is a very good iron, in the bottom of the crucible. I have often seen this phenomenon.

Copper pyrites also yields up its copper when held in bottle glass. To deal with black copper [oxide] or even copper matte [sulphide ore] and produce the purest rose copper it is sufficient to cast it onto a crucible full of bottle glass and leave it at melting heat for a certain time.

It is necessary that glass to be used in assaying should be charged with as much phlogiston as possible. There is no need to fear the volatilization of the metals; the glass gives to the metal oxides only as much of the inflammable principle as is necessary to reduce them. The materials that provide the glass with the largest amounts of phlogiston are clays, lime-rich earths, fusible spar, fullers earth, black-calcined bones and soda from Alicante. Other types of soda do not have the same success because they contain far too much neutral salt, Glauber's salt or rock salt to remove the inflammable principle from the glass.

To obtain a glass fully charged with phlogiston, rapid vitrification is no less necessary than the choice of raw materials; if melting is slow the earthy parts and even the fixed salts lose their inflammable principle much more easily than the glass loses it and there is a much greater loss than when all are combined together in the glass. This glass should therefore be melted using the most violent fire.

Four parts of marl of the type that becomes moderately hard in glowing embers and one part of Alicante soda give a glass very suitable for the assaying of iron ores. In places where marl is rare it may be replaced by a mixture of equal parts of common or brick clay and lime. The only

---

c. See my memoir on the *electrical fluid*.

point needing attention is that if the clay contains a lot of silica sand, like that found widely in parts of the world of ancient creation, it is best to extract it by repeated washing and sedimentation.

Equal parts of clay and gypsum with one fifth of soda from Alicante have always given me an easily melted glass highly charged with phlogiston when melted in a glass furnace; if one adds pure clay it will be almost equally effective as a flux for assaying all ores.

However, the glass that I believe to be the best to ensure the accuracy of the assay of all types of ore is that made from two parts of blue, green, violet, &c. fusible spar and one part of bones calcined in a completely covered crucible, or one part of soda from Alicante. Eight parts of this glass, finely powdered, with one part of finely powdered ore have always given me accurate and rapid assays.

This glass seems to me to fulfil all six of the conditions that we have indicated. It is very fixed in the fire, melts very easily and has a very marked affinity for *matrices* rich in clay, silica, or lime. It is therefore very effective in removing the mineralizers; it is very fluid and it carries a sufficient quantity of phlogiston.

In passing we must note that our procedure presents just one considerable difficulty. As the flux is more fluid and has more energy to vitrify all earths than lead glass, the only crucibles that can resist it are those of glass-porcelain and those of the composition of our new pottery.

We have been able to find a method of decreasing the attack of ordinary crucibles. That is to coat their insides before use with a viscous white glass made from three parts of silica and two of fixed alkali. This glass fills completely even the very smallest pores and the crucibles can then resist the action of our assaying glass for a considerably longer time. An addition of viscous glass, as just described, or even of pure quartz, makes this flux less tractable, less tenuous and less energetic but it is very difficult to determine exactly the most advantageous proportions and I must admit that I have been unable to do so.

I do not claim that I have advanced the art that I have treated here, I have only made some comments. It is to be hoped that those who have more skill and more aptitude will take them up, make the necessary experiments and give us what we at present lack in this regard.

CRITICAL EXAMINATION OF EXPERIMENTS MADE ON SELENITIC AND VITREOUS SPARS

Printed in the
Memoires Litteraires, Critiques &c. 1776

BEFORE the beautiful experiments of Marggraf chemists had only faulty ideas about vitreous and selenitic spars; if they distinguished some by their external characteristics they seemed to confuse their internal properties, attributing the same nature to them all. They seemed to regard them all as fusible and phosphoric minerals.

We owe to the skilled chemist in Berlin the most complete proof that chemistry can provide that Bologna stone [barium sulphate] and all those of the same type, which are very numerous, are selenitic salts, that they are composed of lime and sulphuric acid with a little less than one sixth of clay. He also proved that gypsum [selenite] is of the same nature as Bologna stone, but differing in the proportions of the constituents and by some [auxiliary] materials that are found in one but not the other: that stones of the same type as Bologna stone have about one seventh more lime and about one seventh less sulphuric acid and *vice versa*; also that in these there is never any clay but there is a little more than one sixth of water but the others never yield any. The results of this chemist also show that vitreous spars are of a completely different nature.

Thus errors so ancient and so widely believed are almost always put before great men as respectable truths. Such is the general designation that naturalists and miners give to vitreous and selenitic spars despite the fact that Marggraf has brought into the clearest light the reasons for rejecting that and which he has himself adopted. According to his experiments stones of the Bologna type can never be melted by themselves even in the most violent fire. They can only facilitate the melting of metals from which it is proposed to release the foreign materials present as gangue, as well as others that would not have been decomposed by a fixed alkali, or by phlogiston, or when the gangue of the ore contains clay or has had clay added. Who cannot see that in waiting for these spars to produce the good effects hoped for in the melting of ores, one would lose precious time and that one would have to depend of circumstances rarely coming together? Experience is here perfectly in accord with these principles.

Far from the selenitic spars facilitating the melting of ores, we have repeatedly seen that they retard it and cause great obstacles to arise. It would be difficult to assess the expenses that the error of the miners have occasioned.

They have had a much better opinion of vitreous spar; metallurgy has no more valuable flux. It can be melted by itself in the heat of a glass furnace and one can see proof in our memoir on the false emerald of the Auvergne. Pott, in his *Lithogéognosie* showed that these vitreous spars, however little of lime that one adds to them, they melt more easily. They acquire so great an activity in the melt that we know of no other flux that is so active.

To overcome the harmful effects of the error of the miners it would have been very important for Marggraf to accord the title *fusible* to the selenitic spars. It is very difficult, not to say impossible, for scientists who have studied the arts [only] in their study or laboratory often to avoid confusing errors and prejudices with healthy practice.

One understands that the chemists have given the description *phosphoric* to Bologna stone because it glows in the dark after its calcination with wood charcoal over an open fire. Marggraf himself retained that term for all those of the same nature but this description is not exact because these spars are not themselves phosphoric and they only show this effect when calcined in an open fire with powdered carbon. To avoid stretching the truth and prevent misunderstandings, a correction should be made and these spars called *artificially phosphoric*. However, almost all bodies both natural and manufactured, at least those that are not charged with too great a quantity of water, merit the same description. According to the experiments of M. Beccari and the knowledgable M. Wilson, these bodies, with some preparations are luminous in the dark.[a]

Wilson has made new experiments and discoveries that hold great promise.

Vitreous spars of all colours are phosphorescent [luminescent] by themselves and this property is a characteristic. We believe that we have proved that in our memoir already mentioned. If you simply scatter a powdered vitreous spar onto hot coals it spits and produces a pale green flame; if it is the false emerald it gives blue and the false sapphire provides a very agreeable kind of firework. Thus all types of roughly crushed stones that produce a luminous and coloured trace on hot coals can be regarded as vitreous spars. It is very astonishing that Marggraf did not perceive this property of vitreous spars: he probably never calcined them in sufficiently dim surroundings when he would have seen that as soon as the heat penetrates them they become phosphoric [luminescent] and this phosphorescence ceases as soon as they lose their colour, which happens very quickly.

According to what we have said it seems evident that Marggraf should have refused to give the title *phosphoric* to the selenitic spars and given it exclusively to the vitreous spars and that the former should in mineralogy have only the name of selenitic spars and the latter only that of phosphoric spars. There are those of this latter type in England and France that have crystals so small that they do not appear vitreous [lacking transparency?] and there are some of the first type that do appear to be vitreous.

Two chemists, Scheele and Sage, have appeared to dispute the discovery and nature of fusible spars; they have both made experiments on these stones as varied as curious. No one can today doubt that one can obtain by distillation of phosphoric fusible spars with sulphuric acid, even when cold, an acid which eats away and decomposes glass, makes alkalis

---

a. See series of experiments on phosphors.

caustic, deliquescent, gelatinous, &c. Sweden claims that its chemist has the priority and much of Europe, including France, appears to be persuaded. Is there not a way of ending this disagreement in a manner that is satisfactory even if not agreeable? We believe that it may be found in the details of the experiments of Marggraf. It seems to us that the rigour of the discovery belongs to neither the Swede nor the Frenchman but to the Prussian. The latter, it is said in the account that he gave of his experiments, mixed eight ounces of pulverized calcined false emerald with eight ounces of clear non-fuming oil of vitriol; this mixture was distilled in a graduated heat and gave, when most of the water had gone, a good white sublimate that increased as the fire was was increased and which collected in the neck of the retort, even almost passing into the collecting flask. The first parts to form had the appearance of butter of antimony and melted like that butter when a hot coal was brought near to the neck of the retort but the last to form in the greatest heat did not melt when a hot coal was brought near. When the retort was broken a residue of twelve ounces was found. Marggraf concluded, contrary to all the rules of chemistry and the phenomena of his procedure, that four ounces of oil of vitriol had combined with the spar. The bottom of the retort was found to be pitted with many holes. It does little for the reputation of a great chemist to conclude that this demonstrated the fluxing property of the spar. He evidently confused the fluxing property of the spar with the dissolving power of its acid. Lastly, this chemist confirms that the liquid that had passed into the receiver and the white sublimate in the neck of the retort which had even reached the receiver had a definite smell of sulphur. This obviously arose from the effect of the violent degree of heat used which had made a part of the oil of vitriol combine with a part of the phlogiston of the spar.

 The sublimate, triturated a long time with hot distilled water, eventually dissolved and passed through the filter; it therefore was a salt of fixed alkali. Having then been poured on the liquor, a precipitate formed and slowly settled to the bottom of the vessel; this was therefore not a salt composed of the earth of the spar and sulphuric acid because that precipitate had been promptly formed and abundant. This precipitate, sweetened and dried, also melted, either in the crucible or over hot coals, or at the enameller's lamp and formed a mass similar to porcelain; this therefore was a particular type of neutral salt from which sulphuric acid was excluded. In another place Marggraf says that the spar makes the alkali strongly caustic.

One might tend to suspect that this knowledgable chemist from Berlin must have had particular reasons for not giving the true explanation of these phenomena and the results of his procedure because he saw all the essentials of what was later seen by Scheele and Sage: a liquor that passed into the receiver, a white material that collected in the neck of the retort and even passed into the receiver, erosion or decomposition of the glass undoubtedly by the liquor that was formed, that the white material dissolved in hot water, that the alkalis could precipitate very little, that the precipitate could easily enter into a type of vitrification. What finally appears to me to give the priority of the discovery beyond doubt to Marggraf is that Scheele and Sage followed his procedure exactly; like him they used distillation, like him they used equal parts of false emerald and sulphuric acid, &c. &c. It is possible that Scheele and Sage had no knowledge of the procedure of Marggraf and that they were led to it by lucky chance. The priority is nevertheless due to the chemist from Berlin: the public has gained no less from having very clear ideas of how developments occurred.

OBSERVATIONS ON THE EVAPORATION OF WATER THROWN ONTO MOLTEN GLASS

Printed in the abbé Rozier's
Journal of Physics, May 1778

EXPERIENCE proved that water expands; there is no one who does not believe that it evaporates when put on a fire. Stahl confirms that a bomb full of water put into a fire breaks into fragments, making a frightening explosion. Glass makers, being convinced not only that water evaporates itself when heated but also assists the volatilization of other materials when thrown onto molten glass in the pot, [use it] to dissipate bubbles or the gall that causes them, &c.

The evaporability of water would not be less destitute of all foundation if the phenomenon reported in the *Observations of Physics and Natural History* for last January had any reality and if one could regard as sound the explanation given there. M. Deslandes* (said the author of the note) last year showed the Duke de Rochefoucauld and myself a surprising phenomenon which appeared so extraordinary that it seemed to contradict all that has been written about the properties of water. M. Monet and many other physicists have repeatedly witnessed it during the course of this year: thus it is a fact and an experiment as well authenticated as one could wish.

The Duke de Rochefoucauld, M. Monet and many others, can attest that when the contents of a wooden spoon holding a good glassful of water are thrown into the pot onto the molten plate glass (which is about to be cast) the water remains undisturbed as it falls onto the glass but rolls

---

* Director at Saint-Gobain (1758–89), successor to D'Antic.

around the surface as if it were molten metal, that it does not give off any visible fumes; it quickly takes on a spherical form without the least noise; it appears to take on a red colour similar to that of the pot and glass containing it; that it needs more than three minutes [timed by] a watch in the hand for it to be completely evaporated; that on another occasion M. Deslandes, not being able or not wishing to wait for the water to be completely evaporated, had the molten glass cast on the table to form a plate without there being any explosion.

To explain this phenomenon M. Deslandes said that the sudden evaporation of water only occurs in other circumstances because the surrounding or ambient air being in close contact with the surface of the water gives it, so to speak, wings; but in the present circumstances the extreme heat completely rarefies the air and totally removes it above the surface of the glass and even around the pot so that there cannot be any detonation; on the contrary, not being able to *volatilize* the water takes up a greater degree of heat than it normally would in evaporating; it melts, so to say, and appears in a state which has been ignored until now.

This phenomenon, which has just been described, would assuredly be the most surprising, the most extraordinary that has ever been observed; the explanation offered gives it even more zest, adding much to its eccentricity. It presumes first that the evaporation of water is uniquely due to contact with the ambient air; second, that at the first instant, the complete rarefaction of the air hinders the evaporation of water but that three minutes later it does not prevent it; thirdly that detonations should be completely attributed to the air; fourthly, that water vapour rarefied as much as it can possibly be remains visible; fifthly, that suddenly rarefied water, the expansion of which is not impeded by any obstacle, should detonate, &c.

However, before examining the explanation let us assure ourselves of the reality of the phenomenon, that water can remain on molten glass without evaporating.

As soon as one throws water onto molten glass one does really see globules which resemble shiny drops of mercury on the surface of the melt; these globules can only deceive people having little familiarity with the heat that is necessarily produced in vessels full of a very hot substance. It is easy to convince oneself that these small spheres are hollow and astonishingly fragile; they can easily be collapsed against the walls of the pot, they can be pricked with an iron wire, etc. The workmen know a method, as certain as it is simple, of making visible at close range by eye, under the microscope in the hand, these supposed globules of water. If the glass necessary to make a blown cylinder of sheet glass, or a large bottle is well prepared on the end of a blow pipe and pierced, that is to say a cavity about the size of a goose's egg blown in it, the workman, having taken some water in his mouth, can blow it down the blow pipe then immediately afterwards close the mouthpiece with his thumb. Spectators who do not know this little trick are then very astonished to see the mass of glass expand into a perfectly round globe, sometimes more than three feet in diameter, without being blown or with any other cause; one can see inside this globe very mobile globules that one might take for quicksilver or molten metal. This experiment can be performed on a more or less large scale in all glass houses. It seems to me to prove that water obviously expands and evaporates on and in molten glass.

Having carefully broken this glass globe it is easy to confirm that the globules which roll around its interior are hollow, extremely easily burst and dull coloured; but how are these globules produced? Rectified spirit of wine thrown onto red hot iron behaves in a similar way, see p.281 of the first part of Boerhaave's *Treatise on Fire* (French translation). This problem would be much more difficult than that which we have just resolved.

BIBLIOGRAPHY

Modern Works

Barrelet, J., *La Verrerie en France*, Librairie Larousse, Paris, 1953
Barton, J.L., *Bosc D'Antic and Saint-Gobain*, Glass Technology, **30** 115–6 (1989).
Cable, M., *Threads of Glass*, Glass Technol. **29**, 181–7 (1988): (b) *Bosc D'Antic and Saint-Gobain*, Glass Technol. **30** 116 (1989).
Daviet, J.-P. *Une multinationale à la Française. Saint-Gobain 1665–1989*, pp.334, Paris, Fayard, (1989).
Hamon, M. *Du Soleil à la Terre. Une Histoire de Saint-Gobain*, pp. 214, Paris, Éditions Jean-Claude Lattès, (1989).
Hamon, M. and Perrin, D., *Au cœur du XVIIIe siècle industriel*, pp. 756, Paris, Éditions P.A.U. (1993).
Harris, J.R. *Saint-Gobain and Ravenhead*, pp. 27–70, in *Great Britain and Her World, 1750–1914*, ed. B.M. Rockcliffe, Manchester University Press ISBN 0 7190 0581 7, (1975).
Piganiol, P., *Le Verre, son histoire – sa technique*, Hachette, Paris, pp.261 (1965).
Strathern, P., *Mendeleyev's Dream*, Penguin, ISBN 0-14-028414-1, (2000).

Older cited works traced through the British Library catalogue

Agricola, G. (ca. 1485–1540), *De Re Metallica* (1556); translation by H.C. Hoover and L.C. Hoover, London (1912); Dover reprint.
Agricola, G. *De Ortu et Causis subterraneorum; De Natura Fossilium ...* Basiliae [Basel], H. Frobenium & N. Episcopium (1546, 1558).
Bergman, T.O. (1735–84), *Physical and Chemical essays ...* translated

by Edmund Cullen, 2 vols. London, John Murray; Edinburgh, J & J Fairbairn (1788).

Biringuccio, V. *The Pirotechnica of Vannaccio Biringuccio*, Venice (1540); English translation by C.S. Smith and M.T. Gnudi, New York (1990) ISBN 0-486-26134-4.

Blancourt, H. de, *De l'Art de la Verrerie*, Paris (1697).

Boerhaave, H. (1668–1738), *Elementa Chemiae quae anniverso labore docuit* ... 2 vols Lugduni Batavorum, Isaracum Severinum; (1732).

Cramer, J.A., *Elements of the art of assaying metals. In two parts* translated from the original Latin, London, Tho. Woodward; C. Davis pp. 470, (1741).

Cronstedt, A.F. (1722–65) *An Essay towards a System of Mineralogy* translated from the original Swedish by G. von Engestrom, Edward & Charles Dilly, London (1770).

Dossie, R. *Observations on the Pot-Ash brought from America ... Processes for making Pot-Ash and barillia in North America &c* London (1767).

Ercker, L. *Lazarus Ercker's treatise on ores and assaying*, translated from the German edition of 1580 by A.G. Sisco and C.S. Smith, Chicago, (1951).

Ferchault de Réaumur, R.-A., *L'art de convertir le fer forgé en acier et l'art d'adoucir le fer fondu*, Paris (1722); [English translation, University of Chicago, (1956)].

Gellert, C.E. *Metallurgic Chymistry ... translated from the original German by I.S. [John Seiferth]* London: T. Becket (1776).

Jars, G. the elder, *G.J.'s Metallurgische Reisen zur Untersuchung und Beobachtung der vornehmsten Eisen-Stahl-Blech- und Steikohlen-Werke in Deutschland, Schweden, Norwegen England und Schottland* 4 vols Berlin (1777–85).

Kulbel, -, *Dissertation sur la cause de la fertilité des terres* Bordeaux, (1741).

Marggraf, A.S. (1709–82) *Opuscules chymiques* 2 vols Paris: Vincent (1762).

Merrett, C., (1614–95) *The Art of Glass*, London (1662); *The world's most famous book on Glassmaking* reprint, ed. M. Cable, Society of Glass Technology (2001) ISBN 0 900682 37 X.

Loysel, *L'Art de la Verrerie* Paris (1791).

Nickolls, John, *Remarques sur les avantages et les desavantages de la France et de la Grande- Bretagne, par rapport au commerce ... Traduction de l'anglois.* Dresde (1754).

Plinius secundus, C. *Historia naturalis. The elder Pliny's chapters on chemical subjects*, translated by K.C. Bailey 2 vols. London (1929–32).

Pott, J.H. *Chymische Untersuchungen ... fürnehmlich von der Lithogeognosie oder Erkäntniss und Bearbeitung ... der Stein und Erden* Potsdamm (1746).

Pauw, C. de, *Recherches philosophique sur les Égyptiens et sur les Chinois* (1773).

J.B. and N. de Ville, *Histoire des Plantes de l'Europe*, Lyon, (1680).

Réaumur, R.-A. *see* Ferchault.

Rouelle,-., *Chemical Observations*, J. Medecine, Paris, October (1777).

Sage, B.G. *Des Herrn Sage chemische Untersuchungen verschiedener Mineralian* übersetzt von L.A.G. Schrader Göttingen (1775).

Schlosser, In *Essays and Observations Physical and Literary* [vol. 1 (1754)]; suppl. vol. 13 (1766).

Stahl, G.E. (1660–1734) *Fundam. chymiae dogmaticorationalis et experimentalis* (1732) [an earlier version was published in 1723 and two editions of a Pharmaceutical chemistry appeared in 1721 and 1728.

Strabo, (?63BC–AD24) *The Geography of Strabo, with an English translation by Horace Leonard Jones*, London: William Heinemann; New York G.P. Putnam's Sons. (1917–32).

Swedenborg, E. (1688–1772) *Opera philosophica et Mineralia* 3 vols. Dresdae et Lipsiae [Dresden & Leipzig] (1734).

Wallerius, *Mineralogie, ou description générale des substances du règne mineral* translated from the German by Baron P.H.D. von Holbach, 2 vols Paris (1753).

Woodward, J. (1665–1728) *Essay towards a Natural History of the Earth and Terrestrial Bodies* London: printed by T.W. for Richard Wilkin (1702).

Woulfe, P. (?1727–1803) *Experiments made in order to ascertain the Nature of some Mineral Substances* ... London (1777).

PLANTS USED TO PROVIDE ALKALI

Several plants are mentioned as sources of ash for glass making. Although given Latin names by D'Antic, those names are obsolete. The following are the best modern equivalents that can be traced.

Kali majus cochleato. This is *Salsola soda* L. an erect maritime species referred to in the *Species Plantarum* of 1753. It is not found in Britain but grows on the coasts of the Mediterranean, such as those of Provence, Languedoc and Roussillon. *Salsola kali* of prostrate habit is found in Britain and known simply as saltwort.

Kali majus geniculatum is probably the prickly saltwort, *Salsola kali* L. which is widespread around the coasts of Europe.

Salicor refers to the genus *Salicornia* which includes several species called various types of saltworts in English and which grow in salt marshes. In Europe these are often referred to as *Salicornia europaea* or *S. herbacea*.

Kali de Sicile is probably *Salsola verticillata* Schousboe, which appears to be the only species recorded for Sicily.

Kali à feuille capillaire velue may be *Salsola vermiculata* L. which is well known in the western Mediterranean and southern Portugal. It may refer to *Salsola papillosa*, which is endemic to Almeria or to *salsola webbii*.

Kali à feuille de geneste is *Salsola genistoides* Juss. ex Poiret.

Kali à feuille de tamarisque is *Salsola tamariscifolia* Lag. which was formerly confused with *S. genistoides*.

Absynthe maritime is wormwood, *Artemisia absinthum* L.

fenouil maritime refers to fennel, *Foeniculum piperitum* DC.

Tabac refers to tobacco; *Nicotinia glauca* is well established around the Mediterranean.

Fougère can refer to many ferns or bracken.

Varec may be best translated as sea-wrack but may include a variety of brown seaweeds such as *Fucus* and kelps. *Goëmon* refers to common brown seaweeds.

The book mentioned by D'Antic is probably *Histoire des Plantes de l'Europe* by J.B. and N. de Ville published in Lyon in 1680 and reprinted several times during the next century.

The above information was kindly supplied by Prof. A.J. Willis of the University of Sheffield.

INDEX

Acid, aerial, 10, 168, 206, 211
–, mineral, 94, 103, 106, 218
–, sulphuric, 42–45, 76, 77, 80, 91, 94, 95, 127, 144, 151, 169, 212, 219, 221–223
Agricola, 8, 29, 30, 65, 66, 112, 229
Alexandria, 32
Algae, 71
Alicante, 38, 71, 72, 125, 136, 147, 148, 192, 216, 217
Alkali, fixed, 21, 40, 43, 44, 60, 64–80, 82, 84–86, 89–91, 93, 95–98, 102, 104, 105, 109, 115, 117, 131, 132, 134–136, 140, 141, 148, 150, 166–168, 171–175, 179, 182, 183, 201, 203, 205–207, 209, 211, 212, 217, 220, 222
Allier (river), 159
Alsace, 68, 141, 174, 178
Ambassador, Dutch, 4
America, 3, 33, 172, 176, 193, 230
Anor, 37
Animal glass, 94–96
Annealing, 6, 19, 31, 52, 58, 98, 100–102, 107, 114, 132, 137, 181, 184, 186–189
Antimony, 61, 97, 120, 167, 191, 194, 207, 223
Aprey, 4, 144, 146
Ardennes, 68
Arsenic, 96, 120, 128, 139, 167, 202, 203, 206, 207, 214
Ash (residue), 40, 58, 64, 65, 67, 68, 69, 71, 75, 97, 98, 117, 129, 132, 150, 166, 171, 172, 173, 178, 233
Ash (wood), 139
Aspen, 99, 139

Assaying, 5, 11, 201–203, 207, 214–217, 231
Astronomy, 29
Athénée, 32
Auvergne, 5, 44, 46, 50, 60, 62, 87, 90, 153–155, 157, 159, 161, 167, 194, 215, 220

B

Bark, 97, 98, 139, 172, 173
Batch, 7, 9, 21, 30, 39, 46, 48, 61–66, 69, 72, 76, 78, 80, 82–89, 93, 96–98, 102, 104, 107, 116, 117, 119, 120, 124, 126, 129, 131, 134, 135, 139, 140, 147, 148, 156, 160, 164–167, 174, 177, 182, 183
Baigory, 195
Bar-sur-Aube, 42
Bayel, 85
Bayreuth, 37
Beast, 162
Beccari, 221
Beech, 75, 98, 171, 172, 178, 183
Berlin, 115, 219, 223
Belière, 43, 155
Bergman, 63, 65, 81, 229
Birch, 98, 139, 177
Blancourt, 30, 42, 52, 113, 231
Blisters, 31, 96, 115, 174, 182
Blowpipe, 185
Boëlu, 44, 46, 155
Boerhaave, 9, 31, 227, 230
Bohemia, 35, 38, 40, 55, 57, 85, 100, 132, 174, 181, 182, 192
Bone, 87, 94, 95, 96, 103, 165, 216, 217
Bordet, 46, 154, 155

Bordpré, 154, 155
Bottle glass, 37, 51, 52, 83, 132, 164, 166, 191, 216
Bourbon-L'Archambault, 163
Bourbonnais, 194
Box (wood), 171
Bracken, 67, 173, 233
Brandenburg, 37
Brioude, 154
Brittany, 42, 193
Brunswick, 193
Bubbles, 31, 39, 69, 81, 86, 96, 98, 99, 113–121, 128, 131, 148, 166, 209, 225
Bucket-maker, 193
Burgundy, 39, 42

C

Calcination, 63, 64, 69, 74–78, 93, 96, 105, 106, 174, 177, 178, 204, 214, 220
Canada, 36, 173
Carbon, 10, 46, 80, 94, 95, 96, 98, 125, 126, 135, 139, 168, 174, 201, 204, 205, 207, 209, 220
Carinthia, 195
Carthage, 39, 136, 147, 148
Cementation, 77, 89, 93, 95, 110, 111, 159, 214
Champagne, 42, 85
Chandelier, 34, 35, 36, 37, 58
Charcoal, 74, 77, 135, 139, 140, 207, 220
Cherry, 51, 98, 184
Chimay, 154
China, 109–111, 192
Chloride, 11, 66, 97, 98
Cimey, 44

Clay, 33, 41–52, 57, 59–61, 78, 91, 105, 109, 111, 144–147, 149–152, 154–160, 164, 168, 178, 184, 187, 188, 191, 192, 202, 204, 214–217, 219, 220
Clermont, 154
Cleuzel, 62, 63
Cologne, 46
Cobalt, 80, 81, 194
Copper, 15, 76, 77, 80, 184, 191, 203–205, 207, 214, 216
Cord, 31, 36, 38, 41, 52, 74, 79
Coucy, 37, 192
Cronsted, 88, 232
Crucible, 60, 95, 115, 155, 164, 201, 205, 208, 209, 214, 216, 217, 223
Crusher, 37, 74
Cullet, 82, 83, 86, 102

D

Dannemora, 195
D'Antic, Paul, 2, 3
D'Antic, L.-A.-G., 5
Dauphiné, 192
Decolorizing, 11, 84
Delln, 37
Deslandes, 4, 7, 8, 225, 226
Deux Ponts, 193, 195
Devitrification, 88, 134
Diamond, 100–102, 181, 188
Diderot, 2
Dieuze, 68
Dijon, 16, 80, 103, 123, 143, 153, 161
Diospolis, 32
Distillation, 45, 80, 94, 95, 168, 221, 223
Drulanvaux, 85

Index 237

E
Egypt, 32
Enamel, 33, 88, 93, 107, 120, 128, 143–149, 151, 152, 192, 194, 223
England, 3, 5, 35–37, 58, 110, 193–195, 197, 221, 231
Épinac, 42
Ercker, L., 202, 231
Evaporation, 58, 73, 74, 79, 91, 97, 132, 133, 136, 140, 149, 171, 172, 176, 179, 225, 226

F
Faience, 5, 11, 46, 69, 143–152, 212
Financiers, 22
Fir, 171, 172
Fireclay, 145–147, 151, 152
Flattening, 137, 182, 185–189
Flint, 5, 11, 34–38, 40, 44, 45, 58, 60–63, 75, 78, 82, 88, 90–92, 99, 105, 111, 167, 168, 182, 191–193, 207, 208
Fluid, electric, 92, 123–129, 213
Flux, 20, 39, 65, 66, 70, 75, 78, 83, 85, 86, 89, 95, 104, 105, 106, 108, 129, 147, 148, 153, 156, 160, 166, 195, 201, 202, 204, 205, 211, 214, 215, 217, 220, 222
Foam, 64, 69
Folembray, 192
Forez, 42
Forges, 42
France, 3–5, 27, 34, 37–42, 51, 60, 66, 85, 110, 143, 150, 153, 156, 159, 161, 174, 191, 192, 193, 221, 222, 231
Franconia, 35, 40
Fritting, 83, 97, 148, 150
Furnace, 5, 6, 16, 19, 29–31, 34, 36, 39, 41–60, 73, 78, 83, 85, 86, 91, 93, 96,

97, 99–102, 107, 108, 110–116, 119, 125, 134, 137, 138, 144, 146, 148, 150, 155–157, 159, 164, 174–178, 181–189, 191, 198, 217, 220

G
Gahn, 80
Gall, 31, 62, 64, 65, 69, 81–88, 91, 92, 96–99, 114–120, 124–126, 131–133, 135–141, 147–150, 167, 204, 225
Gangue, 163, 202–207, 214, 220
Ganister, 41
Geller, 202
Germany, 7, 35, 38, 40, 42, 66, 159, 191, 193
Giromagny, 162
Glass, animal, 94–96
– bottle, 37, 51, 52, 83, 132, 164, 166, 191, 216
– crown, 35, 55, 107, 181, 193
– cylinder, 107, 182–186, 227
– green, 19, 33, 37, 39, 61, 72, 75, 84, 91, 96, 126, 132, 140, 161, 164, 165, 191, 192
– lead, 36, 109, 148, 160, 164, 192, 215, 217
– optical, 34, 35, 40, 100, 193
– sheet, 100, 102, 133, 156, 181, 184, 193, 227
Glauber, 207
Gluten, 44–46, 155–157
Gold, 8, 77, 92, 93, 153, 191, 199
Great Britain, 34, 36, 37, 195, 196, 229
Grinding, 19, 20, 134, 152, 201
Grog, 35, 43, 46–50, 52, 149, 150, 154, 188

Index 238

H
Hainault, 37
Hanover, 193
Hazel, 97, 139
Health, 25, 109
Henckel, 30, 65, 66, 84, 94, 113
Herculaneum, 33
Hesse, 153, 159
Holland, 3, 159, 191, 194
Hornbeam, 99, 171, 172
Horsehair, 43, 47, 146, 158
Horse chestnut, 173, 179
Hydrofluoric acid, 163, 168, 212

I
Ipsen, 153, 159
Iron, 8, 11, 15, 17, 21, 42, 45, 47, 50, 51, 54, 56, 61, 73, 75, 76, 77, 79, 80, 81, 96, 104, 109, 110, 111, 118, 120, 128, 137, 145, 176, 177, 178, 183, 185, 186, 187, 188, 189, 193, 194, 196, 203, 206, 207, 214, 216, 227
Italy, 33, 66, 110

J
Japan, 192
Javogue, 44, 46, 154, 155, 158

K
Kali, 70, 71, 132, 232
Kiln, 137, 138
King's Garden, 94
Knor, 133, 136, 138
Kundel, J. 30, 65, 66, 72, 79, 82, 113
120, 143

L
Laboratory, 13, 17, 220
Lambron, 46, 154
Langeac, 62, 154
Langres, 5, 192
Languedoc, 70, 71, 72, 150, 233
Laone, 42
Levant, 33, 66
Lezon, 154
Lille, 143
Lime (chemical), 7, 10, 33, 44, 82, 91, 95, 97, 98, 99, 106, 109, 111, 132, 137, 141, 144, 145, 150, 152, 155, 163, 164, 165, 167, 168, 182, 203, 209, 211, 215, 216, 217, 219, 220
Lime (wood), 139
Limoges, 44, 154, 193
Limousin, 46, 50, 109, 194
Liquor, 85, 90, 91, 94, 117, 222, 223
London, 34, 37, 44, 99, 169, 171, 179, 181, 201
Lorraine, 68, 178, 193
Loubeyrac, 161, 162, 163
Loysel, 231
Luminescence, 128, 163

M
Madagascar, 63
Malachite, 205, 207
Malzieux, 154
Magnesia, 35, 73
Manganese, 11, 34, 65, 72, 80, 81, 84, 86, 90, 93, 98, 124, 125, 126, 133, 135, 136, 139, 140, 165, 182
Manure, 139, 179
Marc, 67
Marl, 42, 45, 144–147, 151, 216
Marsac, 154, 155
Marseille, 194
Marver, 183, 184

Index 239

Memmendots, 162
Méréviels , 42
Merrett, C., 1, 30, 31, 61, 62, 65, 66, 79, 113, 120, 230
Melliona, 192
Metallurgy, 9, 21, 87, 108, 111, 191, 193, 195, 202, 203, 214, 220
Mineralogy, 16, 17, 41, 64, 87, 108, 199, 201, 221, 231
Mirror, 5, 18, 32, 34, 35, 37, 38, 39, 58
Montdebert, 42
Montel de Gelat, 154
Montpellier, 3, 42, 71
Mosaic, 110
Moustier, 143, 192
Murano, 34

N

Nantes, 42, 143
Neri, A. 1, 30, 31, 65, 72, 113, 118, 131
Neronville, 196
Neustad , 34
Nevers, 143, 144, 145, 147, 192
Nollet, abbé, 4, 16, 118, 119, 123, 128
Normandy, 6, 38, 42, 52, 99, 100, 155
Novion, 85
Nuremberg, 35, 193

O

Oak, 75, 99, 129, 171, 172, 179
Opalescence, 11, 131–141, 165, 166
Orleans, 194

P

Painters, 107, 194
Palatinate, 35, 38, 40, 52
Paris, 3, 4, 5, 6, 19, 20, 27, 34, 35, 37, 123, 144, 159, 192
Pays de Foix, 193
Pebrac, 161
Phlogiston, 1, 7, 9, 10, 71, 76, 87, 89, 111, 127–129, 132, 136, 137, 139, 167, 168, 202–207, 209, 210, 212–217, 220, 222
Phoenicia, 33
Phosphorescence, 128, 163, 169, 220
Phosphoric oxide, 63, 94, 95, 103, 106, 168, 210, 213, 219, 220, 221
Piedmont, 84
Pine, 171, 172, 206
Plate glass, 5, 6, 19, 20, 32, 34, 36–39, 52, 54, 99, 100, 101, 109, 112, 114, 115, 124, 126, 132, 133, 134, 136, 137, 138, 156, 161, 193, 225, 226
Poitou, 70
Polishing, 19, 20
Pont-au-Mur, 154
Pontoise, 194
Poplar, 99
Pot, 6, 7, 44, 48, 58–63, 86, 89, 98, 99, 101, 115–118, 120, 125, 132–134, 138–140, 155, 156, 158, 160, 165, 167, 225–226
Potash, 11, 36, 37, 39, 64–67, 69, 70, 74–90, 102, 137, 139, 140, 141, 146, 148, 149, 150, 171–180, 182, 205
Pots, 17, 47, 48, 52, 80, 84, 89, 97, 114, 115, 127, 161, 162, 163, 211, 214, 215, 220, 230
Potter's clay, 45, 144–147, 149, 151
Provence, 70, 233
Prussian blue, 75–78
Puits, 157
Pyrenees, 193, 194
Pyrites, 42, 43, 202, 216

Index 240

Q
Quartz, 44, 45, 49, 50, 60, 62–65, 78, 90, 155, 161, 162, 168, 217

R
Ravenhead, 5, 229
Réaumur, 4, 14, 17, 89, 110, 165, 166, 196, 205, 211, 230, 231
Réfining, 19, 20, 41, 60, 83, 86, 97, 99, 102, 115, 124, 135, 139, 166, 182
Retort, 94, 222, 223
Riom, 154
Rochefoucauld, 225
Rocherta, 65, 66
Romilly, 4, 6
Roland, Eudora, 5
Roland, Mme, 5
Roots, 42, 157, 172, 173
Rouelle, 94–96, 103, 106, 231
Rouen, 143, 192
Roussillon, 193, 133
Royal Society, London, 5, 168, 178, 201
Rupert, Prince, 118, 119

S
Sage, 63–66, 80, 96, 103, 107, 127, 168, 205, 207, 210, 212, 221, 223, 231
Salt, neutral, 64, 71, 85, 98, 114, 126, 148, 150, 171, 203, 204, 205, 206, 215, 216, 223
Salt, sea, 64, 68, 69, 72, 115, 137, 147, 148, 149, 150, 178, 179, 180
Saint Flour, 154
Saint Germain, 46, 154
Saint-Gobain, 2–8, 12, 18, 20, 39, 40, 52, 124, 229
Saint-Quirin, 87
Sainte-Marie aux Mines, 81
Saint-Thiery, 44, 46
Sand, 7, 19, 30, 34, 41–44, 46, 48, 49, 50, 53, 60–64, 70, 72, 78, 82, 84, 85, 86, 89, 90, 93, 98, 104, 109, 111, 115, 117, 125, 129, 134, 141, 143, 144, 146–150, 152, 154, 155, 157, 158, 165, 167, 175, 179, 182, 192, 211, 217
Sandstone, 41, 49, 50, 60–62, 86
Sauxillanges, 44, 46, 154, 155, 156, 157, 158
Saxony, 35, 42, 66, 192, 194
Schlosser, 96, 168, 231
Selenite, 64, 94, 168, 180, 219
Sèvres, 37, 110, 192
Sheet glass, 100, 102, 133, 156, 181, 184, 193, 227
Siberia, 193, 194
Sicily, 8, 70, 192, 233
Sidon, 32
Siege, 48, 53, 54, 58
Silver, 11, 45, 153, 194, 199, 206, 227
Slag, 47, 50, 51, 97, 110, 159, 204, 205, 207
Smears, 69, 114, 131, 135, 136, 138, 167
Smoke, 58, 99, 133, 134, 152, 177
Soda, 7, 19, 20, 38, 64–67, 69–74, 78, 79, 83, 86, 93, 96, 125, 129, 132, 133, 136, 137, 141, 147–150, 166, 192, 216, 217, 233
Soot, 69, 98, 125, 126, 139, 179
Soupe, 196
Spain, 38, 70, 193, 194
Spar, 63, 87, 89, 103, 106, 161, 162, 163, 164, 168, 169, 212, 214, 215, 216, 217, 220, 221, 222, 223

Index

S
Stahl, 9, 127, 225, 230, 231
Steel, 11, 17, 110, 111, 162, 193, 196
Stourbridge, 44
Straw, 42, 175
Styria, 193, 194
Sulphated tartar, 63, 64, 69, 76, 97, 98, 114, 125, 137, 140, 148, 149, 167, 171, 174, 178, 179, 180, 204, 208
Sulphur, 125, 127, 162, 202, 203, 206, 214, 222
Suzi, 42, 216
Sweden, 51, 110, 193, 194, 196, 222
Sweet chestnut, 172, 179

T
Tarn, 3
Tartar, 64, 69, 76, 97, 98, 114, 125, 126, 137, 140, 148, 149, 167, 171, 174, 178, 179, 180, 201, 204, 206
Tax, 37, 193, 194, 197
Thebes, 32
Théric, 192
Thiers, 154
Tiérache, 85
Tobacco, 67, 117, 233
Touraville,
Turgot, 4
Turkey, 33
Tyre, 32
Tyrol, 193

U
Usson, 154

V
Valdenburg, 159
Velay, 155
Viane, 3
Villenrode, 42
Vosges, 85, 162, 193

W
Wallerius, 41, 42, 61, 84, 161, 231
Water, 8, 9, 11, 43–46, 61, 64, 72–74, 76, 82, 83, 91, 94, 95, 97, 104, 116–120, 135, 141, 145, 149, 150, 155, 157, 158, 172, 174–177, 179, 180, 183, 185, 186, 192, 208, 210, 219, 221–223, 225–227
Willow, 99, 139, 171, 172
Wilson, 221
Windsor, 44
Woulfe, 233

Z
Zaffre, 84, 98, 125, 126, 194
Zinc, 76, 77, 80, 81, 213

www.ingramcontent.com/pod-product-compliance
Lightning Source LLC
Chambersburg PA
CBHW051634230426

43669CB00013B/2303